Unbuilding Cities

INSIDE TECHNOLOGY
edited by Wiebe E. Bijker, W. Bernard Carlson, and Trevor Pinch

A list of the series appears on page 275.

Unbuilding Cities

Obduracy in Urban Socio-Technical Change

Anique Hommels

The MIT Press
Cambridge, Massachusetts
London, England

© 2005 Massachusetts Institute of Technology

MIT Press books may be purchased at special quantity discounts for business or sales promotional
use. For information, please email special_sales@mitpress.mit.edu or write to Special Sales
Department, The MIT Press, 55 Hayward Street, Cambridge, MA 02142.

Set in Bembo by The MIT Press. Printed and bound in the United States of America.

Library of Congress Cataloging-in-Publication Data

Hommels, Anique.
Unbuilding cities : obduracy in urban socio-technical change / Anique Hommels.
p. cm. — (Inside technology)
Includes bibliographical references and index.
ISBN 0-262-08340-X (permanent paper)
1. City planning—Netherlands. 2. Urban renewal—Netherlands. I. Title. II. Series.
HT169.N4H647 2006
307.1'216'09492—dc22   2005041686

10 9 8 7 6 5 4 3 2 1

CONTENTS

ACKNOWLEDGEMENTS

Academic work is never produced in isolation; it is always rooted in institutional, intellectual, and social contexts. This is the place to thank the people who, representing these different contexts, helped shape this book.

I benefited most from the numerous discussions I had with Wiebe Bijker and Karin Bijsterveld. They were intensely involved in the project—from my ideas about potential questions and possible cases to detailed discussions of draft chapters. Wiebe and Karin share an inspiring enthusiasm for research in Science, Technology, and Society studies. I am grateful for the confidence they have always expressed in this project and in me, and I surely hope that our collaboration in both research and teaching will continue in the future.

The Faculty of Arts and Culture, and later MERIT/Infonomics, proved to be stimulating environments for my research. I thank all those colleagues who regularly put interesting articles, books, references, or calls for papers in my pigeonhole. I am particularly grateful to the members of the BOTS research group for their willingness to read draft chapters and to provide detailed comments in a fine, supportive atmosphere. Of this group I want to thank Ernst Homburg, Jens Lachmund, Bernike Pasveer, Geert Somsen, Ger Wackers, and Rein de Wilde in particular. I also benefited much from discussions with Ruth Benschop, Marco Goud, David Hamers, Ruud Hendriks, Stine Jensen, Eric Lemmens, Jessica Mesman, Ruth Mourik, Peter Peters, and Jessica Slijkhuis.

The participants of several workshops and conferences in the Netherlands and abroad provided helpful input. The "Technological Futures-Urban Futures" workshop in Durham in 1998 convinced me that bringing together perspectives from urban studies and STS would prove to be a productive venture.

This book could not have been written without the help of many interviewees. Architects, planners, representatives of local activist groups, politicians, consultants, and project managers shared their views, their hopes, and their concerns with me. They made me realize that the research on which this book relies could never have been done within the confines of academia only, that it is necessary to relate academic research to the "real-life" problems of urban practitioners and citizens.

I am also indebted to the Society for the History of Technology for awarding me the 2003 Brooke Hindle Postdoctoral Fellowship, which helped me to turn my dissertation into a book. In the final stages, the editors of the Inside Technology series provided extremely useful and challenging input. I want to thank Sara Meirowitz of The MIT Press, who kindly guided me through the whole process of book production. Ton Brouwers and Paul Bethge meticulously edited my English and made numerous useful suggestions for the improvement of the argumentation.

On the different models of obduracy, see my article "Studying obduracy in the city: Toward a productive fusion between technology studies and urban studies" (*Science, Technology, and Human Values* 30, 2005, no. 3: 323–351). For an earlier version of chapter 2, see "Obduracy and Urban Sociotechnical Change: Changing Plan Hoog Catharijne" (*Urban Affairs Review* 35, 2000: 649–676). I thank Sage Publications for permission to elaborate these two articles.

The writing of this book would never have been so pleasurable without the support and distraction provided by my friends and family. My parents always encouraged and supported me in whatever I did. Their unconditional love and confidence are very important to me. I thank Reinout for his intense friendship and support, which date back a long time. Merlijn, our son, my little wizard, enchants me every day with his curiosity about the world around us and his eagerness to learn.

UNBUILDING CITIES

Obduracy in the City: Three Conceptual Models

Unbuilding[1] Cities

Utrecht, December 1997: The City Council approves plans to demolish a quarter of Hoog Catharijne, a generally despised but commercially successful shopping mall right in the middle of the city's downtown area. It took 10 years of debate and controversy to reach this decision. The indoor mall was planned and built in the 1960s as a part of the Plan Hoog Catharijne, a large-scale redesign of Utrecht's city center. In addition to the mall, the Plan Hoog Catharijne comprised a new railway station, a bus station, new infrastructure, offices, cultural facilities, and apartment buildings, all integrated and interconnected. In the mid 1980s, the negative effects of the Plan Hoog Catharijne became more and more apparent. In 1987 the city initiated a new project aimed at upgrading the area: the Utrecht City Project (UCP). Despite Hoog Catharijne's commercial success, its overall concept began to be generally perceived as outdated and its architecture as ugly; the drug addicts and homeless people who populated the indoor mall in ever-larger numbers damaged its image further. For quite some time, though, it seemed highly unlikely that the mall would ever be touched in the slightest way; it had become accepted as a fact of life. In 1997 it seemed very likely that part of Hoog Catharijne would be demolished. But in 2000 the mall's

Figure 1.1
Drawing of original Plan Hoog Catharijne (1963). Source: Utrechts Archief.

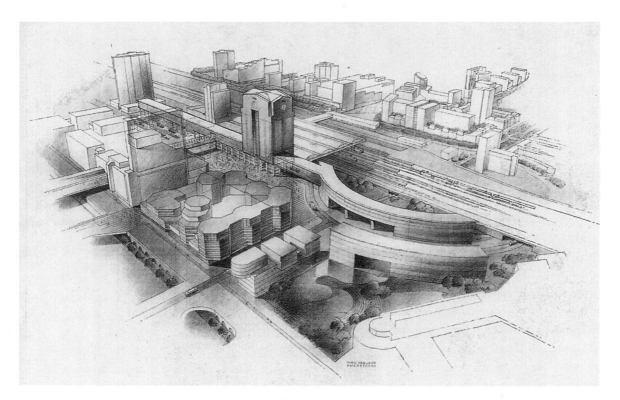

Figure 1.2
Artist's impression of plan for reconstruction of Hoog Catharijne (1995). Source: Projectbureau
UCP.

fortunes turned again. The members of the partnership that had been formed to implement the UCP ended their cooperation, and Leefbaar Utrecht, a new local party whose main goal was to prevent execution of the renewal plans, won some seats on the City Board in local elections. In 2003 a completely new master plan was approved by the City Council, but it is still rather uncertain whether Hoog Catharijne will maintain its obduracy.

Maastricht, October 1998: The Minister of Transportation decides to postpone all planning activities for a major highway reconstruction project in Maastricht until after 2012. A two-kilometer stretch of highway that divides the city is the last part of a highway system that cuts through a densely populated urban area without overpasses.[2] As a result of an extraordinary increase in the number of automobiles since the 1960s, local congestion has increased substantially, especially during peak hours. At the same time, Maastricht's overall accessibility and livability and the safety of its traffic have deteriorated seriously. This stretch of highway is seen as the last bottleneck on the "autoroute du soleil" running from Amsterdam to Genova. Since the early 1960s, when the highway was built, engineers, politicians, and citizens have vied to change and adapt it. Building a tunnel has been considered from the very beginning, but so far neither that solution nor any other has been implemented. Despite all efforts at altering the highway's design to allow through traffic to avoid the city, the road has maintained its obduracy. After the national elections of 2002, the City Board put much effort into getting the commitment of the newly elected politicians. In 2003, the national government promised that a tunnel will be built after 2007.

Amsterdam, July 1999: During a visit to the Bijlmermeer, one of Amsterdam's suburban districts, I witness the demolition of a huge multi-level parking garage and a shopping center. Since 1992 a number of apartment buildings have been torn down in this part of the city. According to the present plans, only ten of the original thirty apartment complexes will survive. The Bijlmermeer (commonly abbreviated to "Bijlmer") was built in the 1960s and the 1970s according to a functionalist design. Its high-rise buildings and spacious apartments were

Figure 1.3
The highway that cuts through Maastricht (1968). Source: Gemeentearchief Maastricht.
© Gemeentearchief Maastricht. Photograph by J. Naseman.

Figure 1.4
Congestion on the highway through Maastricht (ca. 1985). Source: Gemeentearchief
Maastricht. © Fotopersbureau P. Mellaart.

originally intended for middle-class Dutch families. Quite soon, however, it became clear that such families were not attracted to the Bijlmermeer, and it became a refuge for minorities and immigrants, who had begun to enter the country in larger numbers. By the 1980s, the Bijlmermeer, plagued by unemployment and crime, had become one of the most criticized urban districts in the Netherlands. For a time, changes in its basic design were not seriously considered. But in 1992, after years of discussions involving the city government, the neighborhood council, housing corporations, community workers, and residents, it was decided that a rigorous redesign of the Bijlmermeer would have to be undertaken in order to solve its urgent social problems and the severe financial problems of the housing corporation. This meant a radical break with the original concept, as over the years many of the huge apartment buildings had been replaced by single-family homes. Because there appeared to be less crime and vandalism in the low-rise areas, it was decided in 1999 that an even more rigorous renewal of the Bijlmermeer would be needed: half of the remaining apartment buildings would be demolished and replaced with low-rise structures. Meanwhile, a small number of Bijlmermeer residents try to preserve part of their living environment by establishing a "Bijlmer Museum" to conserve and display the original designs.

These three vignettes illustrate the central theme of this book: the confrontation between ongoing attempts to change cities (sometimes even recent additions or innovations) and the obduracy of existing urban structures. The book deals with the clash between new ideas about urban development and the opinions and policies embedded in the urban structures that are already in place. It addresses the unexpected or unforeseen societal developments that gradually give rise to questioning of existing urban configurations. It is about urban design and attempts to renew it—a process in which the stakes often are so high that years of planning, debate, and controversy may result in no changes at all in some cases, whereas in other cases urban reconfiguration may eventually result.

The projects described in the vignettes above are indicative of the boom in planning activities and large-scale spatial redesign efforts that occurred in the 1990s in the Netherlands.[3] Remarkably, projects built in the 1960s have already

Figure 1.5
Characteristic spatial features of Bijlmermeer (April 1976). Source: Archive Dienst Wonen Amsterdam. © Dienst Wonen Amsterdam.

Figure 1.6
Demolition of Koningshoef apartment building (April 1999). Source: Archive Dienst Wonen
Amsterdam. © Dienst Wonen Amsterdam.

been remodeled, "facelifted," and even demolished.[4] Because trends in architecture and ideas about the role of the automobile in cities and about the spatial planning of urban areas changed profoundly, many large-scale planning interventions of the 1960s were written off as failures in the 1990s.

In the Netherlands, a small country where building space is perceived to be scarce,[5] the tension between new developments and old structures is felt most dramatically in attempts to redesign established urban spaces. In many cases, of course, existing structures must be removed or destroyed to facilitate new development. Redesign plans play an important role in public debates because such plans may have far-reaching consequences for people's daily lives. As the journalist H. J. A. Hofland noted, in the Netherlands "national conflicts are about space."[6] As a result, many plans remain controversial for long periods before a compromise is reached, and even comparatively minor redesign projects may turn into money-gobbling and time-consuming affairs.

The wide range of planning initiatives and activities in the Netherlands confirms the image of city building as an ongoing process: cities are being built and rebuilt all the time; they are never finished but always under construction, always being realized. Many plans to redesign urban space assume that the existing urban configuration is almost infinitely malleable. The urban historian Josef Konvitz claimed that "nothing may look less likely to change in a radical way than the status quo in city building, but nothing else may be more likely."[7] It seems counterintuitive to change cities, but nevertheless they change continuously. But despite the fact that cities are considered dynamic and flexible spaces, numerous examples illustrate that it is very difficult to radically alter a city's design: once in place, urban structures become fixed, obdurate, securely anchored in their own history and in the histories of the surrounding structures. Objects and facilities that define urban space tend to coagulate into an amorphous whole. As a consequence, urban artifacts that are remnants of earlier planning decisions whose logic is no longer applicable may prove to be annoying obstacles to urban innovation.

It is not my intention to provide tools for judging the desirability of either changing urban structures or maintaining them, to argue for changing cities, or to plead for their preservation. My study is more theoretically inspired. It deals with the confrontation between "new" plans for urban development and "old" urban structures. How can cities be adapted to accommodate newly conceived ideas and policies? Why do urban structures maintain their obduracy despite efforts at urban innovation? How do special-interest groups and politicians deploy strategies to change what seemed solidly in place or to hold on to what has become contested? My study focuses on this tension between the dynamics and malleability of urban space, on the one hand, and its hardness and obduracy, on the other.

The most obvious examples of obduracy in urban contexts involve buildings, facilities, or structures that have never been contested; they have simply "been there," noticed by few or never in conflict with other potential uses of their locations. To me these sites are not the most interesting ones. More interesting are the urban sites and structures that have been subjected to "unbuilding activities"— locations or elements of cities that are disputed or contested, or that at some point were included in redesign plans. The obduracy of urban structures is "tested" in efforts to "unbuild" them. Debates about redesigning cities or parts of cities are infused with questions about the flexibility or obduracy of a city's elements. To be sure, not all the existing elements of a city are equally contestable, nor is the intensity of particular debates the same in every case or at every stage of the planning process. Yet by concentrating on attempts at city renewal we should be able to gain insight into the circumstances under which urban change eventually becomes possible.

I will concentrate on these tensions between obduracy and change in three Dutch urban redesign projects: the facelift of a city center as part of the Utrecht City Project, the reconstruction of a highway in Maastricht, and the spatial renewal of the Bijlmermeer. In choosing projects to study, I strove to find a balance between homogeneity and heterogeneity: the case studies should have enough in common to be comparable, but they should be different enough to allow for interesting generalizations. For the sake of heterogeneity, I chose a downtown area (Hoog Catharijne), an urban highway, and a suburban district. Moreover, the three empirical examples are unique in themselves. Hoog

---

Catharijne was the first large-scale reconstruction of a downtown space in the Netherlands in which a city cooperated with a private building company; later it became an example for many other similar projects in other Dutch cities. The highway through Maastricht is one of the last stretches of the Dutch highway system to run through a densely populated part of a city. The fact that the highway is fully encapsulated by the city, which itself is locked into the Maas River Valley, raises extremely challenging redesign problems. The Bijlmermeer is one of the few city districts in the Netherlands to have been built according to strictly applied modernist design principles; it has large, identical high-rise apartment buildings, a separation of traffic flows, and huge public car-free zones.[8]

Despite the uniqueness of each of the projects and the obvious differences between them, there are some similarities between them that make it interesting to consider them together. The first of these similarities is that all these projects were originally planned and built between the late 1950s and the early 1970s, and none of the redesign or reconstruction efforts have been completed. The second is that all three case studies are characterized by several comparable themes: the role of infrastructure in relation to urban planning, issues of social safety, criminality, and livability, housing and demolition issues, mobility, the role of the government and national policies, public-private partnerships, and so forth. In addition, all three projects are closely intertwined with Dutch middle-class life. The stretch of highway that cuts through Maastricht is regulated by a series of traffic signals, and this interrupts the fluid passage of the annual exodus of Dutch families, who in their cars and vans have to get past Maastricht to reach the south of France. Hoog Catharijne, for some the national symbol of drabness, certainly did not become more appealing with time; moreover, it began to be increasingly frequented by drunks, addicts, and homeless people. The Bijlmermeer, built for neat middle-class families, became socially degraded and rife with crime and drug activity. The fact that the projects I focus on are all Dutch makes it easier to compare them and allows me to be more precise and specific about the particular Dutch cultural and socio-political context within which these projects figure.

Foreigners visiting the Netherlands always think that the country is planned "from above" because its spatial structures look harmonious, neat, and orderly, but scholars have pointed out that planning in the Netherlands is somewhere between "chaotic planning and planned chaos."[9] Planning takes place "at several levels, without central command."[10] Scholars argue that the harmonious and ordered impression made by the spatial structures of the Netherlands is attributable to the strong role of the public sector in the development, exploitation, and maintenance of the physical environment. A crucial characteristic of Dutch spatial planning is that the local municipalities have forceful land-use policies.[11] This contrasts with the situation in many other European countries and in the United States. The anthropologist Constance Perin concludes that in the United States "the system of land use, its beliefs and practices is in important aspects a *national* one, due to federal government standards and regulations, to the freedom to invest anywhere with the same expectations of return and to the employment mobility of a group having a high level of skill in finance, development, architecture, planning and management."[12]

Urban historians have also observed cultural differences in urban planning between European and American cities. Germany and France have strong traditions of neatly planned infrastructure, leaving less room for technological innovations and resulting in a more "rigid" urban space. The design and evolution of American cities tends to be influenced more by outside factors, on account of which American cities are also more receptive to radical infrastructure innovation than European cities.[13] In general, market forces are having a more profound influence in American cities than in Dutch ones. Americans have a "deep-rooted ideological antipathy to government intervention in urban and regional development."[14]

Despite these international differences between planning cultures and political systems, I am convinced that the specific examples of unbuilding processes I have chosen for this book are relevant beyond the Netherlands. What is happening in the Netherlands is very much like what is happening in other parts of the world. Obduracy and change in urban environments are as problematic in Glasgow, Paris, and Boston as in Maastricht, Utrecht, and Amsterdam. Although Boston's Central Artery/Tunnel Project was much larger in scale than the Maastricht highway, the processes of negotiation and discussion about alternative designs were similar in the two cases. In the Netherlands we have the Bijlmermeer, but several countries have their own landmarks of failed modernism. The Pruitt-Igoe complex in St. Louis and Hutcheson Town "c" in Glasgow are similar projects where demolition was chosen as the ultimate solution to the problems. Furthermore, the developments around Hoog Catharijne's shopping mall and railway station should be considered in the context of the 1980s, when several European cities (including Lille, Stockholm, and Basel) started to make plans to adapt the areas around their railway stations to accommodate high-speed trains. Huge expansions of railways stations, including the introduction of new concepts of public transportation and increasing amounts of office space, were planned and implemented in several European cities.[15]

However, this study is not meant as an international comparison of urban renewal projects. The three Dutch case studies allow for a detailed empirical analysis of three unbuilding processes. What happened when the Utrecht City Project was confronted with the obduracy of Hoog Catharijne? How did it affect the UCP's ideas and priorities? To what extent did the actors succeed in changing the seemingly obdurate design of Hoog Catharijne? Why did the existing structures of the highway that cuts through Maastricht maintain their obduracy despite all efforts to construct a tunnel? Why was the demolition of the Bijlmermeer (or a portion of it) initially out of the question, and why have radical changes of its original plan nevertheless occurred since 1992? How and to what extent can the proponents of the Bijlmer Museum succeed in their struggle to conserving the original planning structures, at least in part?

Empirical analysis of the tension between obduracy and change in these urban redesign projects is the focus of my chapters 2–4. But my concerns are not solely empirical. My book also has two interrelated theoretical aims. First, I intend to apply STS (Science, Technology and Society) concepts to the study of cities. In this book I will view cities as large socio-technological artifacts. The city is the result of human interactions, constructions, and representations. It is an ensemble that includes the material (roads, buildings, bridges, tunnels, transportation facilities, communication systems) and the immaterial (legal regulations, institutions, communities). Thus, I aim to show that cities can be fruitfully analyzed with the conceptual tools of technology studies that were earlier applied to other technological artifacts. Second, I aim to contribute to the theoretical understanding of the role of obduracy in socio-technical change. The issue of obduracy and change informs one of the current theoretical debates in STS studies. On the basis of the case studies, I seek to refine various theoretical concerns about the role of obduracy in urban socio-technical change. In contrast to earlier STS studies that focused on the early stages of technological development, I will concentrate on the process that involves negotiations and attempts at undoing the socio-technical status quo in a city, changing the taken-for-grantedness of its reality, and making its obduracy flexible. It is clear that obduracy and urban change are major concerns both for urban scholars and for practitioners such as architects and urban planners. Therefore, it is interesting to use obduracy as a focal point in exploring the question how STS and urban studies might benefit from each other. In the final section of this chapter, I will propose three models of obduracy that emerge from STS. By applying these three models to cases of urban unbuilding, I aim to show their complementary analytic utility.

VIEWING CITIES AS TECHNOLOGICAL ARTIFACTS

In studying the tension between socio-technical change and urban renewal, I will take the city as a basic unit of analysis. A city, conceptualized as a technological artifact,[16] consists of a wide array of erratic and heterogeneous elements that we must take into account if we are to begin to understand its complexity more comprehensively. Of course, the city as a technological artifact is (like other technologies) never purely technical—it is a "seamless web"[17] of material and social elements. Furthermore, I view planning as a process of socio-technical change. In order to understand the development (and the redevelopment) of cities, it is necessary not only to understand technological processes but also to look at social processes and interactions taking place in the urban context. Therefore, in this book, the city is conceptualized as a socio-technical artifact— a perspective that will be developed, refined, and brought into sharper focus in the course of the book.

Surprisingly, STS studies have paid little attention to the city per se or to the city as a strategic research site for the study of other issues.[18] Moreover, the few studies that have been done mostly address technologies *in* the city, rather than the city *as* a technology. In a 1997 article in *Technology and Culture*, Julie Johnson-McGrath noted that "only a handful of book-length works have addressed the shape and shaping of urban technology."[19] In another article published the same year, Simon Guy, Stephen Graham, and Simon Marvin observe that "since Lewis Mumford's path-breaking books addressing the wider links between . . . technologies and urban history . . . only a few urban historians have attempted to understand how cities and technical networks co-evolve."[20]

Studies of large technical systems focus on specific technological networks, such as electricity, transport, and waste networks. The city as such is hardly mentioned. The authors appear mainly interested in how these networks were built and how the various actors took part in the development of new technologies. Technological systems and networks serve as the basic category of analysis in these studies. Rather than being the focus, the city functions as a mere locus.[21] Although the founding father of the systems approach in the history of technology, Thomas Hughes, situated his analysis of electric power systems in cities,[22] many of his followers have not elaborated on this theme.[23]

Recently, however, some studies have been published that are most appropriately situated at the intersection of STS and urban studies.[24] The urban scholars Stephen Graham and Simon Marvin (1996, 2001), for instance, made worthwhile efforts to explicitly integrate STS theories (Social Construction of Technology, Actor-Network Theory, Large Technological Systems) with urban perspectives such as urban political economy and "relational theories of contemporary cities."[25] A similar line of argumentation has been advanced recently by some American sociologists and philosophers, including Thomas Gieryn, Steven Moore, and David Brain, who point to the importance of space and place in the sociological research agenda and who argue convincingly that such STS concepts as interpretative flexibility, actor networks, and black boxing can be fruitfully applied in analyses of the built environment. Gieryn, Moore, and Brain argue that, because STS concepts pay attention to both the social shaping of technology (or, here, spatial artifacts) and the simultaneous technological shaping of society, they have more to offer than the traditional sociological concepts that can be found in the works of Giddens, Bourdieu, Harvey, and Foucault.[26] In a similar vein, the historians of technology Mikael Hård and Marcus Stippak (2003) ask how historians of technology could contribute to urban history. They argue for a "broad, cross-disciplinary approach" in which more attention is paid to the role of engineers in urban debates and to the relationship between city images (as reproduced in art and literature) and technological form.

Having noted that in STS attention for the city is almost absent, I add that it is also remarkable that the material aspects of the city seem to be neglected in the dominant theoretical perspectives of urban studies. Probably as a result of this, the specific issue of obduracy in urban change does not seem to belong to the subject matter of urban studies at all. For a long time, urban ecology was the dominant approach in research on the social dynamics of cities. Urban ecology was developed by Chicago School sociologists at the beginning of the twentieth century. In the field of human ecology, spatial relations are the analytical basis for understanding urban systems. The Chicago School ecologists Robert Park, Ernest Burgess, and Louis Wirth see the city as a kind of social organism. They explain urban development through a "biotic" determinism, a kind of Social Darwinism of space. This means that competition for the best strategic location (the one where profits can be most easily maximized) is the main underlying mechanism guiding urban development.[27] These ideas can be understood in the American context of the early twentieth century, when laissez-faire economics and privatization dominated the socio-political scene.[28]

Insofar as technology or material factors figure in their analyses, Chicago School ecologists focus on communication and transportation technologies.[29] Example: "Modern methods of transportation and communication—the electric railway, the automobile, the telephone, and the radio—have silently and rapidly changed in recent years the social and industrial organization of the modern city."[30] For urban ecologists, technology seems to be an exogenous force that has a strong influence on the city, and for this reason they have been accused of technological determinism.[31]

Neither the role of technology nor the role of resistance to change in cities has been thoroughly conceptualized by Chicago School ecologists, though Ernest Burgess points to some "complications" that may occur in the expansion of cities. In his influential "concentric zones" model, Burgess developed the idea that cities consisted of a number of circular regions, the central business district being the innermost zone, followed by a zone for industry, one for "working men," a residential zone, and a zone for commuters. Burgess admits that this ideal model for urban expansion might be hampered by existing physical, ecological, or social structures, such as railroad lines, rivers, factories, or "the resistance of communities to invasion."[32] But although it is recognized that urban expansion is not a trouble-free process, urban ecologists do not conceptualize the underlying mechanisms and causes.

In the mid 1970s, the scientific orientation of urban studies started to shift. Urban ecology was criticized for its market-driven economic determinism and for its exclusion of political and cultural factors. New approaches to the city originated from an interdisciplinary mix of neo-Marxism, urban geography, and political economy. In their analysis of the city as a growth machine,[33] the urban scholars John Logan and Harvey Molotch adapted urban ecological points of view to include political developments and cultural institutions.[34] In the neo-Marxist urban geography perspective represented by scholars such as David Harvey, cities are seen as mirrors of the contradictions and flaws of the capitalist system. Capitalism, according to Harvey, is inherently expansory—its goal is maximum mobility of goods, capital, water, energy, and information products. But in a world where infrastructure networks must be embedded in space this is impossible. Cities, the basic units of production and consumption, are fixed and embedded in space.[35] This means that capitalist expansion can be hampered by the fixity of urban structures. Again we see recognition among urban scholars that urban structures can be difficult to change. Nevertheless, Harvey's approach is arguably not very sophisticated in regard to the complexity of the relations between the material and the social and on the role of obduracy in urban change. One reason for this is that Harvey's theory is too monocausal: it relates everything to capitalism, thereby neglecting other relevant factors.

We may thus conclude that something important is missing from the perspectives that dominate the field of urban studies. Arguably, urban scholars do not have the proper conceptual tools to analyze the complexities of relationships between the social and cultural and the material in processes of urban change. In general, they fail to appreciate the significance of obduracy in urban development.

In this book I propose to focus explicitly on obduracy as a major stumbling block in processes of urban socio-technical change. Throughout the book, I aim to show how focusing on obduracy makes it possible to look at urban form and processes of change in a new and different way. I shall do so by elaborating and extending three different models of obduracy. In doing so, I will be refuting four "commonsense" explanations of urban obduracy.

The first "obvious" explanation for urban obduracy is that change is too expensive. Many people tend to think that urban obduracy is directly caused by a lack of money. Throughout the book, I will show that the situation is often more complex than that. Financial considerations can be a reason to keep things as they are, but they can also be a reason to start unbuilding processes. Moreover, financial stakes are not the single cause of obduracy, and they are often inextricably linked to other interests. And, as the sociologist Donald MacKenzie has argued, costs are socially constructed.[36] This implies that "costs" arise from interactive negotiation and calculation processes in which various (non-monetary) values also figure—that "costs" are not a factor in themselves.

The second "commonsense" view holds that there is no agreement on what should be done. As I will show, conflicts of interests are indeed crucial in many unbuilding processes. The process of seeking consensus on what should be done is often very time consuming. But even when consensus is reached, that is no guarantee that urban structures will become malleable immediately. There are other, more complex reasons for the obduracy of urban structures. I will analyze the complexity of the mechanisms involved in interaction processes around urban structures without reducing them to mere conflicts of interest.

Third, it is often claimed that stasis in urban development can be explained by the fact that powerful voices want things left as they are. This claim starts from a rather monolithic idea of power. A careful study of unbuilding processes shows that rarely is it the case that a single actor "has the power" to keep things as they are (or to change things). I favor a "relational" conception of power that emphasizes the "attribution" of power to certain actors, rather than actors' being "in possession" of power.[37] This implies that power balances may change frequently during unbuilding processes, and that hence opinions about what should be changed also vary.

The fourth "commonsense" notion of urban obduracy states that urban structures are difficult to change for material reasons. None of the "commonsense" notions mentioned so far takes the role of materiality into account. However, some approaches in urban history and architecture can be criticized for their *exclusive* focus on the material aspects of obduracy.[38] I dispute the notion of "material obduracy"—the idea that cities, buildings, or infrastructures have inherent technical properties that resist change. Although it may not be technically difficult to demolish an apartment building or to adapt a city highway, such structures may nevertheless prove very obdurate in some "immaterial" sense.[39] Obduracy, then, cannot be explained only by reference to the solidity of concrete and the physical properties of technologies; a wide range of cultural factors come into play.

In contrast to the four "commonsense" accounts of obduracy that focus on single-factor explanations, my study will reveal the complexity and heterogeneity of processes of urban socio-technical development. Urban innovation, conceived as a mode of socio-technical change, involves a laborious, time-consuming, and precarious process marked by a delicate interplay of various social, technical, cultural, and economic factors. By focusing on only one or two of these factors, urban change and redevelopment can be understood only poorly and incompletely. By concentrating on the actors' ideals, assumptions, and cultural values, it can be demonstrated how cities are shaped and how specific ideas are always built into urban design.

It is my contention that STS research has something to contribute to urban studies with respect to the conceptualization of the myriad relations between the social and the material in cities, and with respect to the specific issue of obduracy in urban change. Since one of the major goals of this book is to specifically contribute to the theoretical understanding of the role of obduracy in urban socio-technical change, I will now address how recent conceptualizations of obduracy can help to improve our understanding of this phenomenon. One of the theoretical implications of viewing the city as a technological artifact is that it becomes possible and productive to analyze it with the same conceptual STS tools that are applied to other technologies, such as bicycles, transport systems, or refrigerators. Focusing on obduracy enables a new and different way of looking at urban form and process.

OBDURACY OF TECHNOLOGY: THREE CONCEPTUAL MODELS

The three conceptualizations and explanations of technology's obduracy presented here have roots in technology studies and in urban studies (urban history, history of architecture, geography). Each conceptual model emphasizes different aspects of obduracy, or foregrounds other explanatory mechanisms in the constitution of obduracy. My aim in this chapter is not to argue which view of obduracy is preferable, but to bring out the complexities of the three conceptions in terms of the issues and questions they address (or fail to address). This specifically means that I will focus on the set of concepts and metaphors used in the various views of obduracy, the explanatory mechanisms they rely on, their disciplinary backgrounds or intellectual traditions, and the types of explanations. I will present the three models as "ideal types," which means that I emphasize the distinctions between them instead of the similarities. It is important to keep in mind that the three models are meant as heuristics for the analysis of obduracy in socio-technical change, rather than as accurate empirical descriptions. In later chapters I will discuss the usefulness of these categories when applied to empirical studies of unbuilding processes. In this confrontation with empirical studies, it becomes possible to analyze their relevance, whether they require adjustment, and what we gain or lose by applying them.

Dominant Frames

The category of dominant frames consists of conceptions of technology's obduracy that focus on the roles and strategies of actors involved in the design of technological artifacts. The constraints posed by the socio-technical frameworks within which they operate will be addressed in particular. The concepts of this category apply to situations in which planners, architects, engineers, technology users, or other groups are constrained by fixed ways of thinking and interacting. As a result, it becomes difficult to bring about changes that fall outside the scope of this particular way of thinking. The concepts in this category are generally used to analyze the *design and use* of specific technological artifacts. As an *interactionist* conception of obduracy, this category highlights the struggle for dominance between groups of actors with diverging views and opinions. In relation to specific technological artifacts, examples of this conception of obduracy include Wiebe Bijker's "technological frame," Michael Gorman and W. Bernard Carlson's "mental models," and Cliff Ellis's notion of "professional worldviews." The concepts of "technological frame" and "professional worldviews" have also been applied to planning. Specifically, the concepts in this category highlight the significance of users (or "relevant social groups") and inventors when it comes to explaining technology's obduracy.

Bijker developed his concept of the "technological frame"[40] in the context of the SCOT (Social Construction of Technology) model.[41] This model defines the obduracy of a technological artifact as a stage in the artifact's development. In the early 1980s, Bijker and Trevor Pinch formulated the outlines of their sociological model of technological development. They distinguished three stages in the analysis of a technological artifact. In the first stage, the "interpretative flexibility" of an artifact has to be analyzed. Bijker and Pinch argue that an artifact's technological development should be described from the viewpoints of various "relevant social groups," because, typically, members of various social groups look differently at an artifact and attribute different meanings to it ("interpretative flexibility"). The second stage consists of analyzing the artifact's stabilization. During the interactions within and between these social groups, one meaning will eventually become dominant: the artifact's interpretative flexibility decreases, its meaning becomes more stable, and finally it will have a single dominant meaning. This "closure" of the artifact's interpretative flexibility implies that its meaning will be quite fixed for a period of time.[42] This fixity of meaning results in technology's obduracy. As Bijker puts it, "previous meaning attributions limit the flexibility of later ones, structures are built up, artifacts stabilize, and ensembles become more obdurate."[43] The third stage in the analysis of an artifact's social construction involves relating "the content of a technological artifact to the wider sociopolitical milieu."[44]

Bijker's concept of the "technological frame" is particularly relevant to the analysis of obduracy. A technological frame is built up during interactions among relevant social groups. It may consist of goals, problems, problem-solving strategies, standards, current theories, design methods, testing procedures, tacit knowledge, user practices, and so forth.[45] For the analysis of obduracy, it is important to consider the role of artifacts as "exemplars." After closure, an artifact becomes part of an technological frame as a "exemplary artifact":

> An artifact in the role of exemplar (that is, after closure, when it is part of a technological frame) has become obdurate. The relevant social groups have, in building up the technological frame, invested so much in the artifact that its meaning has become quite fixed— it cannot be changed easily, and it forms part of a hardened network of practices, theories and social institutions. From this time on it may indeed happen that, naively spoken, the artifact "determines" social development.[46]

Besides analyzing the role of artifacts as exemplars, it is also important to analyze for whom a technological artifact is obdurate and for whom it is not. An actor with high inclusion in a particular technological frame thinks and interacts very much in terms of that technological frame. It is difficult for such an actor to think of alternative technological designs. This may be referred to as "closed-in" obduracy. "Closed-out" obduracy is possible too. This occurs when actors have little involvement with a particular technological frame—when they have "low inclusion." For them, the technology presents a "take it or leave it" choice. Seeing no possibilities for variation within a technological frame, they are left with the choice of either accepting it or abandoning it. In this way, according to Bijker, an artifact can be obdurate in terms of having one fixed meaning or in terms of enabling and constraining interactions and ways of thinking.

Although cities are more complex than singular technological artifacts, Eduardo Aibar and Wiebe Bijker suggest that the SCOT approach is also applicable to more complex and heterogeneous socio-technical ensembles, such as planning projects. They analyze the controversies around the Cerdà Plan for the extension of Barcelona between 1854 and 1860.[47] They consider planning "as a form of technology, and the city as a kind of artifact."[48] Taking the SCOT model as their theoretical framework, they analyze the interactions between social groups and their negotiations concerning the extension issue, and they describe how technological frames were formed during these interactions. The technological frames consisted of the problems that were considered important by the relevant social groups, the various solutions to these problems, and the extension plans they proposed. In the course of the events, two rival technological frames came into being: the "engineers' frame" and the "architects' frame." Aibar and Bijker describe the controversy in terms of opposing technological frames that try to become dominant. They argue that where there is no single dominant technological frame an "amortization of vested interests"[49] generally occurs. This is what happened in Barcelona. Aibar and Bijker show how the final layout of the city "got the mobility and easy traffic attributes from the engineers' frame, while hierarchy and high density of buildings were achievements of the architects' frame."[50]

The "technological frame" concept bears a resemblance to the "technological paradigm" notion developed by the economist Giovanni Dosi. According to Dosi, technological development follows a certain "technological trajectory." A technological trajectory is the direction of change within a "technological paradigm"—that is, as an "'outlook,' a set of procedures, a definition of the 'relevant' problems and of the specific knowledge related to their solution" (Dosi 1982:

148). Dosi claims that trajectories are mainly selected on technological and economic criteria, and that, once established, they can acquire momentum. Whereas Dosi emphasizes economic and institutional factors in the construction of a technological paradigm, other concepts encompass more factors, such as cognitive factors, rules, and expectations. A crucial difference between Bijker's "technological frame" and Dosi's "technological paradigm" is that Bijker does not link frames or paradigms with certain technological trajectories.[51]

Although Gorman and Carlson's "mental model" resembles Bijker's "technological frame," they make a distinction between mental models, mechanical representations, and problem-solving strategies or heuristics. Mechanical representations are very precise images of technological artifacts, whereas mental models are often more diffuse, cognitive ideas. Mental models especially address the inventor or designer. Technological frames also apply to other social groups involved in the development of technological artifacts. Moreover, "technological frame" is a broader concept, since mental models mainly consist of inventors' ideas about the future working of artifacts. Gorman and Carlson emphasize that mental models "are shaped by the inventors in response to social and economic pressures as well as personal preferences."[52]

Scholars who have studied processes of urban change advance a similar view. The historian of planning Cliff Ellis, for instance, has looked at the role of "professional worldviews" in the process of American city planning, in particular the design of urban freeways between 1930 and 1970.[53] He argues that, on account of differences in professional training, members of the various professional groups involved in freeway planning (architects, engineers, urban planners, and landscape architects) held different worldviews, which in turn led to their proposal of different design solutions: "The involved professionals used different ideas and images to advance their goals: intellectual tools acquired through education, professional socialization, and daily practice. Professional worldviews shaped the styles of research, the generation of alternatives, and the presentation of proposals to the wider public."[54] Highway engineers, for example, tried to simplify the problem of highway design and make it calculable by developing engineering standards and using computer models. Land-use planners divided the city into urban zones that had to be defined in legal terms and in terms of the activities that could be performed in various parts of the city. Urban designers interpreted the city as a combination of three-dimensional structures imbued with symbolic meanings. According to Ellis, these worldviews, "as embodied in methodologies, recurring solutions, standards, habitual ways of framing a problem,"[55] are difficult to ignore, since they are closely related to the professional's urge for (intellectual) influence and a good reputation.

As my discussion suggests, the various approaches all emphasize the constraining role of frames—ways of thinking and interacting including values, professional conventions, views of the world, typical solutions, problem definitions, and so on—for specific groups of actors. When certain ways of thinking have been built up around an artifact, it becomes difficult to ignore them, let alone change them. Implicit in these approaches is the assumption that, because certain ways of thinking are narrow in focus or difficult to adapt, the technology involved will become obdurate or will have limited flexibility. This means that obduracy, instead of being caused by material factors alone, is the result of interactions between social groups—interactions that are constrained by specific ways of thinking.

Embeddedness

Within STS, technology is often conceptualized as part of a greater whole. Thus, technological artifacts are not analyzed in isolation, but as part of a larger system, network, or ensemble. STS scholars argue that society plays a crucial role in the shaping of technology and that, conversely, technological developments have an important effect on society; they observe, in other words, a "co-evolution" of technological and societal developments. Applied to the built environment, this idea of co-evolution highlights that building cities implies the shaping of society, or that "civil engineering is also social engineering."[56] At the same time, utopian ideals, cultural values, economic considerations, and power relations are built into the physical structure of cities; there is always a "social shaping of technology" at work.[57] Thus, cities are not purely technical constructs; rather, they are a "seamless web" of material and social elements. In the most radical interpretation, the metaphor of the seamless web suggests that the "social" and the "technical" are two sides of the same coin: the technical is socially constructed and the social is technically constructed.[58]

In this approach, the obduracy of technology is related to technology's embeddedness in socio-technical systems, actor networks, or socio-technical ensembles.[59] In this respect, Thomas Hughes argues that the building of a system is accompanied by fewer difficulties when it has not yet become linked to politics, economics, or other value systems.[60] This category involves a *relational* conception of obduracy: because the elements of a network are closely interrelated, the changing of one element requires the adaptation of other elements. The extent to which an artifact has become embedded determines its resistance to efforts aimed at changing it. Such efforts may be prompted by usage, societal change, economic demands, zoning schemes, legal regulations, newly developed policies, and so forth.

The actor-network theorists Michel Callon, Bruno Latour, Madeleine Akrich, John Law, and Annemarie Mol describe technological development as a process in which more and more social and material elements become linked in a network.[61] They investigate attempts by actors to stabilize that network. But the larger and more intricate a network becomes, the more difficult it will be to reverse its reality. In this way, a slowly evolving order becomes irreversible.[62]

Latour gives a clear example of how a network became more obdurate and less reversible. He describes the late-nineteenth-century controversy between the city of Paris and a number of major private railroad companies concerning subway construction.[63] The socialist city government was looking for a way to guarantee that the railroad companies could not take command of the subway system if a right-wing party were to win the city elections in the future. It found a solution in having subway tunnels built that were narrower than the railway companies' smallest coaches. As the subway network expanded, its design became less and less reversible. The obduracy of the subway network became obvious when after 70 years the railroad companies and the subway companies wanted to link their networks. The engineers who were hired to enlarge a number of tunnels were essentially asked to undo what had been decided earlier. "What could have been reversed by election seventy years ago," Latour concludes. "had to be reversed at higher cost. Each association made by the socialist municipality with earth, concrete, and stones had to be unmade, stone after stone, shovel of earth after shovel of earth."[64]

Implicit in these constructivist views of technological development is a movement from flexibility to inflexibility: technology gradually stabilizes and becomes obdurate. Constructivist work argues that, typically, a socio-technical ensemble or system becomes more rigid and less flexible. According to actor-network theorists, it is equally possible to distinguish elements with varying degrees of malleability *within* a single network.[65] Law stresses that the social should not be privileged: "Other factors—natural, economic, technical—may be more obdurate than the social and may resist the best efforts . . . to reshape them."[66] Callon argues that the possibility of changing a network depends on testing the capacity of the various entities that make up the actor network to resist transformation.[67]

With its emphasis on the networked character of socio-technology, the concept of embeddedness seems particularly suitable for the analysis of cities. As Graham and Marvin remark, "the fundamentally networked character of modern urbanism . . . is perhaps its single dominant characteristic."[68] That some elements of a socio-technical network remain obdurate while other elements change—an idea raised by actor-network theorists—has also been mentioned by urban geographers. David Harvey, for instance, argues that the tension between fixity and mobility in urban space is an important issue: "We know that the built environment is long-lived, difficult to alter, spatially immobile and often absorbent of large, lumpy investments."[69] Harvey argues that there are inherent tensions in capitalism between "fixity" and the need for "motion," mobility, and global circulation of information, money, capital, and so on. Infrastructure networks are so crucial for the reproduction and development of capitalism "because they link multiple spaces and times together."[70] Harvey makes a distinction between infrastructure networks that are "highly" embedded in space and networks that are less embedded. Transport networks are highly embedded because the capital these networks embody consists of pipes, cables, roads, and so on that form the physical structures of modern cities. This brings risks with it: "This inflexibility means that sunk infrastructures go on to present problems later in further rounds of restructuring. . . . Crises emerge when older infrastructure networks which are embedded in space, become barriers to later rounds of capitalist accumulation."[71] According to Graham and Marvin, water and waste networks are highly embedded infrastructures, energy networks are "medium embedded," and telecommunication networks are "high to low embedded."[72] The main explanation for embeddedness in this perspective—and here it differs from actor-network theory—is that these infrastructures embody heavy investments and capital that are literally sunk in specific locations.

Stewart Brand makes a similar point when analyzing the adaptability of buildings.[73] Brand makes a distinction between the various layers of buildings that have different life cycles and that change at different paces. His 6-S scheme (table 1.1) differentiates between the slowest rate of change, which applies to the "site" or geographical setting of a building and which may last forever, and faster changes that occur in other layers. Air conditioning systems, for instance, are usually replaced at intervals of 7–15 years. The "stuff" in a building, its furniture, changes most rapidly, on a monthly or even daily basis. Brand suggests that the main reasons for changing a building are related to new styles, especially with regard to the building's exterior, the need for technical maintenance or repair, technological developments, and the obsolescence of systems in the building.

Table 1.1   Life cycles of buildings, ranked by length (longest at top). Based on Brand 1994: 13.

Site	Geographical setting, urban location ("eternal")
Structure	Foundation and load-bearing elements (30–300 years)
Skin	Exterior surface (now 20 years)
Services	Air conditioning systems, elevators, communications wiring, electrical wiring, etc. (7–15 years)
Space plan	Interior layout (walls, ceilings, floors, doors) (3–30 years)
Stuff	Furniture (chairs, desks, phones, lamps, kitchen appliances, etc.)

In sum, "embeddedness" refers to the difficulty of changing elements of socio-technical ensembles that have become closely intertwined. Changing one element may have consequences for the whole ensemble. Obduracy is no intrinsic property of technologies but can only be understood in the context of its ties to other elements within a network. It is possible to differentiate between degrees of obduracy of different elements in a network or system without assuming a priori that social elements are more obdurate than technical ones. In this model, materiality has a different position than in the category of "frames." Because most of the concepts within this category stem from the actor-network tradition, human and non-human "actants" in a network are analyzed more symmetrically.

Persistent Traditions

The category of persistent traditions comprises conceptions of obduracy that address the idea that earlier choices and decisions keep influencing the development of a technology. Because of this focus on the longer-term persistence of traditions in socio-technical change, I call this category *enduring*. The notions of trajectories, path dependency, momentum, archetypes, and city-building regimes embody this conception of obduracy. The crucial difference with the concepts discussed earlier is that they are less focused on interactions in local contexts than the other two models of obduracy: long-term, structural developments that transcend local contexts and interactions get more attention in this approach than in the previous two. The concepts discussed here are less focused on interactions in local contexts than the previous two models of obduracy. One of the potential disadvantages of the frames approach, for instance, is that it is always focused on groups and always emphasizing the local level. This makes it difficult to point at wider "contextual" or structural factors in the construction of obduracy.[74] Generally, the notions within the category of persistent traditions put more emphasis on the wider cultural context in the explanation of obduracy in cities.

Hughes (1987) has argued that the socially constructed features that became embedded in technical systems in the early stages of their development can have lasting effects. His metaphor of momentum highlights the role of trajectories in patterned technological development and can be used to describe the problems of changing large technological systems during certain stages in its development: "The systematic interaction of men, ideas, and institutions, both technical and non-technical, led to the development of a supersystem—a socio-technical one—with mass, movement and direction. An apt metaphor for this movement is 'momentum.'" (Hughes 1983: 140)

When systems are expanding, they acquire momentum, or "dynamic inertia." Hughes (1994) positions his concept of momentum between the two extremes of technological determinism and social constructivism. When a system has acquired momentum, this means that in that phase the system tends to resist change. Young, developing technological systems are more receptive to social and cultural influences than older systems, which, in turn, affect their environment more. This does not imply that a system in a phase of momentum develops autonomously. As Staudenmaier (1985: 154) remarks, "the momentum model understands the very dynamics of technological change as the result of some technical design embodied within a culture."

It is the emphasis on a long-term cultural context that makes this form of obduracy different from the other categories of obduracy. Hughes emphasizes how the supportive cultural context of a specific electricity-supply system (the "polyphase" system) contributed to the system's momentum in the 1890s. At first, manufacturers reinforced the system's momentum by investing in resources, labor, and factories to produce the equipment necessary for its functioning. Later, educational institutions contributed to the system's development by teaching students the skills needed to operate it. These practices were further spread and consolidated by professional journals. After this, research institutes were established to solve the system's "critical problems" (Hughes 1983). All these factors added to the system's momentum.

With its emphasis on the role of trajectories in patterned technological development, the concept of momentum resembles that of "path dependency," an influential conceptual notion developed in evolutionary economics. Path dependency refers to the idea that past events keep influencing the developmental path or trajectory of a technology. Path dependency develops over a longer period of time and suggests that "local, short-term contingencies can exercise lasting effects" (MacKenzie and Wajcman 1999b: 20). The QWERTY keyboard is a well-known example of patterned technological development. Crude notions of path dependency and trajectories as developing according to an internal, "natural" logic have correctly been criticized by STS scholars, who emphasize the contingent and fluid character of technological development. (See e.g. Pinch 2001 and MacKenzie and Wajcman 1999.) An important difference between the notions of path dependency and trajectories and Hughes's concept of momentum is that the former do not pay any attention to cultural factors. For these two reasons (its neglect of contingency in technological development and its lack of attention to cultural factors), the concept of path dependency seems less relevant to my study. Recently, however, some more sophisticated approaches of path dependency have been developed that focus on processes of "path creation" and "path destruction."[75] These approaches fit better in the general constructivist line of thinking I propose here.

Combining elements of Hughes's systems approach and elements of urban regime analysis, the historians of technology Anders Gullberg and Arne Kaijser (1998) introduced the notion of city-building regimes to explain morphological change in urban contexts.[76] Gullberg and Kaijser consider it a disadvantage that the Large Technical Systems approach focuses on only one technical system, since they are interested in the interactions between different infrastructure systems in cities. A city-building regime consists of "a set of actors and the configuration of co-ordinating mechanisms among them which produce the major changes in the landscapes of buildings and networks in a specific city region at a given point of time."[77] Coordination is mediated by regulatory systems in the city (legal and organizational instruments) and the "political culture" (which includes "more subtle" historically grown behavioral rules and conventions). Using this approach, Gullberg and Kaijser try to explain patterns of urban morphological change. They rightly criticize "macro-studies" that describe the development of cities as evolving from "walking cities" to the "tramway city" to the "automotive city." They criticize such an approach for its technological determinism and for its oversimplification (local and national particularities are not considered relevant). Gullberg and Kaijser applied their approach to the postwar development of Stockholm, where they identified three subsequent regimes: the municipal multi-family housing regime (1945–1979), the private single-family housing regime (1970–1985), and the commercial building regime (1985–1995). For example, the first regime was characterized by a strong hierarchical coordination and network coordination in which the municipality and private partners played a crucial role. It mostly produced multi-family houses. Despite tensions and conflicts within this regime, it was stable and extremely dominant.

I have already mentioned the explicit importance attached to cultural factors in Hughes's explanation of a system's momentum. The historian of technology Rosalind Williams gives an interesting cultural interpretation of persistent traditions in *Notes on the Underground* (1990). By analyzing artificial underground worlds as an "enduring archetype," Williams shows how literary traditions from all over the world have always expressed a concern with the underground, which suggests the persistence of the opposition between surface and depth in our thought. Present-day developments in planning and architecture, particularly the trend to build under the ground, to construct tunnels and subways, and to hide less attractive urban functions, can be understood in relation to the work of the nineteenth-century novelist H. G. Wells. In *The Time Machine* (1895), Wells wrote about an underground world inhabited by the Morlocks, who operated machines and utilities, and an above-ground paradise of nature and leisure inhabited by the Eloi. Williams shows that this tradition of "putting the less glamorous aspects of civilization underground" reverberates in the work of twentieth-century architects.[78] In their urge to deal with overpopulation and with space-consuming distribution networks, roads, central heating infrastructure, and factories, they have turned their gaze to the underground world, so that the surface may still be available for the more pleasurable aspects of life (leisure, recreation, parks, housing, schools, and so forth). Such traditions can be enduring in the sense that they are likely to keep influencing choices and decisions of large groups of people.

Another example of the role of persistent traditions in planning is cited by Sally Kitt Chappell (1989). In her study of designs of railway stations in American cities between the 1890s and the 1930s, she notes the importance of four "arche-typal designs" for large railway stations: New York's Pennsylvania Station (1902–1910) and Grand Central Terminal (1903–1913), the Terminal Station built for the 1893 World's Columbian Exposition in Chicago, and Washington's Union Station (1903–1907). These archetypes were based on existing railway sta-tions, and according to Kitt Chappell their influence on American architects is clearly noticeable in the architectural design of large stations built thereafter. Kitt Chappell points out that these stations belonged to the French Beaux-Arts tra-dition, characterized by monumental features. She explains how "the larger con-cepts behind each station have in some measure persisted."[79] The shared visions of different social groups (architects, engineers, public officials, railroad managers) continued to influence the design of major railroad stations. The emphasis on "archetypes" and "shared visions" makes this analysis fit into the category of per-sistent traditions rather than the category of frames, since in the latter category there is more attention to the differences between groups[80] and less to the struc-tural, cultural, and symbolic factors in the obduracy of urban structures.

In the analysis of obduracy, focusing on the persistence of decisions involv-ing the design and building of urban technologies may be quite useful. This model stresses the long-term effect of such decisions on socio-technology. In contrast to the "interactionist" conceptions of obduracy that were discussed under "frames," however, the conceptions described in this section do not focus on specific social groups that interact in local contexts. Instead, the emphasis is on the role of collectively shared rules and values that transcend local contexts, on culturally rooted traditions that derive their strength from the fact that they are shared by many people. The "technological frame" concept, for instance, allows for actors who have different degrees of inclusion in different frames, but the conceptions of "irreversibility," "path dependency," and "archetype" suggest a more comprehensive or pervasive quality of technological artifacts. It will be evi-dent that this broader cultural conception of obduracy that focuses on persistent traditions enables a further operationalization of the linkages between urban technology and the wider cultural context. The main contrast with the category of "relational" conceptions of obduracy discussed under "embeddedness" lies in this category's emphasis on longer-term continuities, whereas conceptions of embeddedness do not specifically focus on such patterns and long-term (cultural) traditions.

Comparison of the Three Approaches of Obduracy

The models highlight different yet equally relevant interpretations of the phenomenon of obduracy. The three categories are schematically represented in table 1.2. When it comes to our understanding of obduracy, then, the various disciplines—architecture, history of planning, geography, history, sociology of technology—offer similar or interconnected conceptual tools. The use of concepts related to "paradigms" is apparent in both STS and planning history; consider, for

Table 1.2   Three models of obduracy.

	Dominant frames	Embeddedness	Persistent traditions
Explanatory mechanisms	Obduracy explained by constrained ways of thinking and interacting	Obduracy explained by close interconnectedness of social and technical elements	Obduracy explained by long-term persistence of traditions
Concepts and metaphors	Technological frames; paradigms; mental models; professional worldviews	Actor networks; irreversibility; fixity and mobility of space	Momentum; trajectories; path dependence; city-building regimes; archetypes
Disciplinary background or intellectual tradition	Social Construction of Technology; history of planning	Actor-Network Theory; urban geography	Large technical systems approach; history of technology; urban history; evolutionary economics
Type of explanation	Interactionist conception of obduracy	Relational conception of obduracy	Enduring conception of obduracy

instance, the concepts of "technological frame" and "mental model" in STS and the concept of "professional worldviews" in the history of planning. Moreover, there have already been attempts to apply concepts originally developed for the analysis of other technical systems or artifacts to the analysis of cities, of city planning, or of urban artifacts. (Examples include Aibar and Bijker's analysis of Barcelona using SCOT and Latour's story about the subway tunnels in Paris.)

If obduracy is mainly associated with dominant frames, we deal with an *interactionist* conception according to which obduracy can be the result of interactions between various groups of actors. The interactions between the actors are structured and often constrained by the meanings and values they attribute to technologies. In contrast to interactionist conceptions, explanations of obduracy in terms of embeddedness and persistent traditions no longer take social groups as a starting point. Embeddedness involves a *relational* conception of obduracy: it can be explained by the interrelatedness of heterogeneous elements in a sociotechnical ensemble. Obduracy may be the direct result of, for instance, tight relations between the various material and non-material elements. An explanation of obduracy in terms of persistent traditions, my third category, differs from the others because of its more enduring character—its focus on longer-term processes that are deeply rooted in culture at large and that, depending on the specific tradition or pattern, may vary only slightly. A clear difference with the category of frames is its focus on collectively shared rules and values that transcend specific groups and local contexts. Whereas concepts in the category of frames highlight the differences between social groups, a focus on persistent traditions reveals the similarities, what is shared among groups: no group or single actor can easily escape from influential and lasting traditions.

There are also substantial differences between the conceptual frameworks as such, even within STS.[81] The STS approaches discussed—Actor-Network Theory (ANT), Large Technical Systems (LTS), and Social Construction of Technology (SCOT)—originate in different theoretical traditions. ANT grew out of semiotics; LTS is an offshoot of the history of technology and business history; SCOT has roots in symbolic interactionism. Although all three approaches rely on the "seamless web" metaphor as a starting point for research, ANT differs from the others by not accepting a fundamental distinction between human and non-human actors. ANT theorists embrace the "principle of generalized symmetry," which means that the same theoretical framework should be applied to the analysis of human and non-human actors. SCOT emphasizes interactions between "relevant social groups" and the meanings they give to a certain technological artifact. The concept of "technological frame" was introduced to avoid social reductionism.[82] At the same time, it is clear that "materiality" assumes different guises depending on which of the three models one chooses. ANT gives greater weight to material things; SCOT gives greater explanatory weight to social groups. The persistent-traditions model pays less attention to technology and does not emphasize the role of social groups but focuses more on long-term structural developments.

Furthermore, the wide range of "units of analysis" and "research sites" in the various approaches deserves mention. Technological frames and mental models are related to artifacts or technical objects; irreversibility is related to actor networks. How, then, are the three broad categories of conceptions of obduracy that I identified to be "translated" into a useful apparatus for approaching the issue of obduracy in processes of urban socio-technical change?

As I noted earlier, my interest in obduracy is motivated mainly by my concern for efforts that are aimed at reshaping urban technology. None of the concepts discussed above is entirely appropriate for analyzing such efforts. There are two reasons for this: (1) Some of the concepts are related to technological objects and are thus not specifically focused on the analysis of processes of socio-technical change in the city. (2) Most of the concepts discussed address the initial shaping of technology rather than its redesign in the context of urban renewal. Nevertheless, it would of course be a great mistake to reject these conceptions of obduracy altogether. At least some of the concepts discussed have already proven their usefulness in analyzing processes of urban redesign. That other concepts have not yet been applied to the city does not mean that it is impossible or unproductive to do so. By integrating elements of the three conceptions that have been shown to be fruitful in previous analyses, I focus my argument on those elements that I find particularly useful for the analysis of obduracy in cities:

———

- The frames model emphasizes obduracy in design processes. Studying obduracy in urban redesign involves identifying the actors involved in local planning processes and "unbuilding" activities and analyzing their potentially conflicting ways of thinking.

- Embeddedness emphasizes the interrelatedness of human and non-human elements in an urban socio-technical ensemble. This notion nicely captures the heterogeneity of cities: streets, buildings, distribution networks, development plans, politicians, and pressure groups together constitute the large, complex socio-technical ensemble that the city is. In cities, infrastructure, laws and regulations, traffic schemes, usage, urban policies, and planning structures are closely interconnected. In specific circumstances, this can result in the obduracy of individual elements or of the ensemble as a whole.

- A focus on persistent traditions highlights how cultural and collective traditions that persist over a longer period of time and transcend local contexts and group interactions contribute to the obduracy of urban structures. For example, long-term, long-standing traditions of architecture or of planning play a major role in the constitution of the obduracy or malleability of a city's parts.

Apart from this theoretical exploration of conceptions of obduracy in urban socio-technical change, a confrontation between these theoretical conceptions and my empirical case studies is needed in order to refine the conceptions and gain insight into the tensions between obduracy and change in urban redesign projects, and to elaborate on how STS can contribute to studying cities. In the following chapters, I will analyze case studies of the tensions between obduracy and change in three ongoing urban redesign projects in the Netherlands: the redesign of Hoog Catharijne as part of the Utrecht City Project, the highway reconstruction in Maastricht, and the spatial renewal of the Bijlmermeer. I rely on the three models of obduracy discussed above as ways of exploring the explanatory power and specificity of these conceptions in these case studies.

Attempting to apply the three models makes their relevance clear and shows the extent to which they need to be adjusted.

The first goal of this book is to analyze the tension between obduracy and change in three major urban redesign projects in the Netherlands, covering the period between the 1960s and the 1990s. The second goal is to make a specific contribution to the theoretical understanding of the role of obduracy in urban socio-technical change. The first goal has strong historical overtones; the second has a decidedly theoretical orientation. The third goal is to bring the city into the limelight of Science, Technology, and Society studies and to introduce STS to urban scholars. Conceptualizing the city as a socio-technical artifact, I will try to find out to what extent STS concepts can be useful to investigate processes of urban change. The final aim of this book is to contribute to a productive fusion between STS and studies of the city.

Attempts to Change Hoog Catharijne

Hoog Catharijne, a large urban redesign project in the Dutch city of Utrecht, was planned in the 1960s. The project radically altered the look of the downtown district, some parts of which go back to the medieval period. The plan was meant to revitalize the city's economy and comprised a modern shopping mall, apartment buildings, offices, a new railway station and bus station, and a complete reconstruction of the area's infrastructure. For the Netherlands, Hoog Catharijne was a huge project, but in the United States indoor shopping malls were already a trend in those days. In the 1950s, shopping areas become more and more enclosed and "indoor." The number of shopping malls grew fast in the 1960s, and the first mega-projects were initiated in the 1970s.[1] Southdale, near Minneapolis, opened in 1956, was the first indoor shopping mall. In his study of the development of the Dutch shopping mall, the historian of architecture Dion Kooijman compares the design of Hoog Catharijne to the situation in Minneapolis: "Hoog Catharijne is a combination of an elevated urban structure that reminds us of the superstructures (so-called skyways) in the downtown area of Minneapolis and a traffic centre *avant la lettre*."[2]

Despite the plan's ambitious character, many residents of Utrecht regretted Hoog Catharijne even before the complex was officially opened.[3] The project's overall concept was generally perceived as rigid and outdated, its architecture as outright ugly. In 1970, a group of concerned residents founded an activist committee; among other things, it organized a protest rally attended by about 4,000 people on the day of the official opening of the complex in June 1973. Because of its design and sheer size, Hoog Catharijne soon became a pre-eminent symbol of capitalism in the public mind. The following quotation is from a newspaper story published the day before the opening of Hoog Catharijne: "New and very recent is the objection that Hoog Catharijne is a symbol of a society that is directed at consumption and production. . . . There is indeed a growing notion that the continuous desire for more, better, bigger and more beautiful leads us to the abyss that is so realistically described in the Club of Rome's report."[4] Moreover, critics argued, Hoog Catharijne lacked the social function of meeting place inner cities should have. Many saw its buildings as overly functionalist, and their dismal architectural quality was hardly disputed.[5]

When the complex was close to completion, many felt that it already belonged to a bygone era, and the inflexibility of the overall design left little room for future adaptation to newer views: "It is observed that while views about the city's central district are changing, they cannot be realized anymore because of Hoog Catharijne's rigidity. Resistance and irritation are on the rise, and Utrecht's residents increasingly consider Bredero[6] a threat to local living circumstances rather than viewing him as the builder of the city's new future."[7] There was a widespread feeling among critics at the time of Hoog Catharijne's completion that the project's design would not allow the incorporation of new views about city renewal at any point in the future. They emphasized that the sheer size of the complex left no opportunities for alternative projects to leave their mark on the downtown district. Moreover, the complex was designed to expand; it had a built-in tendency to spread, and this would further erode the city's livability. One commentator referred to it as a "lump" and a "tumor" that was growing every day.[8] In the newspapers Hoog Catharijne was portrayed as "immovable," "a dead moloch," "not very flexible," "a massive concrete lump," and "a concrete tumor."[9]

The poor image of Hoog Catharijne was often tied to its major building material, concrete.[10] Seeing mass and immobility as intrinsic properties of

Figure 2.1
Protests against Hoog Catharijne (1973). Source: Utrechts Archief. © Fotopersbureau 't Sticht.

concrete, people seemed to blame the material for the obdurate character of the complex.[11] Bredero, the construction company, was held responsible for Hoog Catharijne's disappointing appearance. The derisive expression "Bredero Beton" (Bredero Concrete) was heard, and the builder's director, Jan de Vries, was called "Jantje Beton."[12]

Others, however, espoused a different opinion of Hoog Catharijne's design. K. F. G. Spruit, one of Hoog Catharijne's architects, emphasized that it was certainly possible to change existing city centers and Hoog Catharijne in particular, claiming that its newer sections had been designed differently and that a variety of materials had been used.[13] Spruit expressed his astonishment about the protests and suggested they were not so much aimed at the use of concrete as a construction material as at "Hoog Catharijne as a large-scale company."[14] Spruit, designer of an office building for a large insurance company (AMEV), compares the public reactions to his building with the controversy about Hoog Catharijne: "Consider the example of the AMEV building. Is anyone talking about AMEV concrete? No, but they are talking about 'Hoog Catharijne concrete' or 'Bredero concrete.' And do you realize that the very same material was used in both cases!"[15] This exclusive focus on the material as a decisive or intrinsic property of a structure's obduracy provides a good example of the commonsensical notion of "material obduracy." Views like those of Spruit suggest that actors may have divergent positions on the obduracy of particular urban structures, even if a huge project like Hoog Catharijne is involved.

In the meantime, in the United States as well as in the Netherlands, the disadavantages of shopping malls had become clearer: high energy costs, a repetition of architectural styles, a high degree of standardization that neglected local identities, and the destruction of traditional urban patterns.[16] Although during the 1970s the residents of Utrecht saw no other possibility than to take their city's changed inner reality for granted in all its concrete ugliness, in the mid 1980s several negative effects of the Plan Hoog Catharijne came to the fore. As a consequence, the city leadership initiated a new project to upgrade Hoog Catharijne: the Utrecht City Project (UCP). Ideas about city planning, infrastructure, traffic circulation, and social safety had changed profoundly in the intervening years. How did these new ideas as reflected in the UCP affect the obdurate structures of Hoog Catharijne? Why did the project's design maintain its obduracy despite all the efforts to rebuild it? How did the actors eventually, if only temporarily, succeed in changing the seemingly unassailable presence of Hoog Catharijne?

CAST IN CONCRETE: THE PLAN HOOG CATHARIJNE

The Plan Hoog Catharijne was developed by Empeo, a subsidiary of the large construction company Bredero. It was generally viewed as a solution for many of the problems Utrecht was facing in those days.[17] Because of a housing shortage, the city needed more space for its growing population.[18] After World War II, the city government, motivated by progressive ideals, articulated ambitious plans for expanding and improving the city. One of the goals was to increase Utrecht's economic capacity.[19]

The City Board put much effort into enhancing the accessibility of the downtown district. The increasing number of automobiles had created congestion, a problem the city already acknowledged in the 1950s. Many felt that the city's infrastructure had to be adapted to accommodate this increase of cars.[20] The city government invited M. E. Feuchtinger, a German expert on traffic circulation, to design a new traffic plan for Utrecht. In 1958, Feuchtinger proposed to construct a peripheral road around the city center. This meant that some canals would have to be paved over. In the late 1950s, Utrecht, like many other Dutch cities, was challenged by the need for huge urban reconstruction projects that involved the building of new peripheral roads and the demolition of urban neighborhoods or parts thereof.[21] It was mostly the increased traffic density that served as a legitimation for filling in canals and rearranging the existing architecture of the city.[22]

Feuchtinger's city renewal proposals would radically change downtown Utrecht. Not surprisingly, they met with opposition. The City Council decided that the planning aspects of the Plan Feuchtinger had to be investigated. It invited the architect J. A. Kuiper to carry out the research. Since Utrecht's leadership foresaw a substantial expansion of the service sector, it wanted to double the amount of office space.[23] Because of the city's population growth, more retail space was needed as well. The main question in the ensuing debates involved the location of the new office buildings, wholesale businesses, and shopping districts. Kuiper presented his plan in 1962. He argued for a de-concentration of the downtown activities over a larger part of the city. Kuiper and other proponents of de-concentration held that unsolvable traffic and parking problems would be caused by locating most of the business and commercial activity in the downtown area. Kuiper's reasoning, however, failed to convince the city leadership.[24]

At the same time, the Dutch National Railway Company, Nederlandse Spoorwegen (NS), was increasingly struggling with a lack of space in downtown Utrecht, the location of its main offices as well as of the city's Central Station, the busiest and most centrally located railway station of the Netherlands. In addition to building a larger railway station, NS wanted to increase the amount of parking space near the station.[25] At the request of NS, Empeo studied the possibilities of building parking garages near the railway station. At first, only the area directly surrounding the railway station was taken into account, but at a later stage this study gave rise to a plan to reconstruct the whole station area. On March 20, 1962, the city leadership and representatives of NS, Empeo, and the building company Bredero discussed a proposal for an integrated reconstruction of the railway station area and its immediate vicinity. Representatives of NS and of the City Board reacted enthusiastically.[26] They gave Empeo six months to prepare an extensive redesign plan.

Empeo put together a multi-disciplinary team consisting of a planning consultant, a social scientist, a traffic expert, a technician, and a construction manager. They developed a plan in consultation with experts of NS and the city. The first Plan Hoog Catharijne was published in October 1962. Based on the results of the investigations of this team and referring to broader discussions concerning the location of business activities in relation to downtown districts, Empeo proposed a concentration of business and commercial activities. In its view, decentralization would have huge disadvantages for shops and offices that had to move to the city's outskirts. Moreover, this would produce, in effect, more car traffic. Apart from that, Empeo claimed that the Plan Hoog Catharijne harmonized well in many ways with the plans of Feuchtinger and Kuiper.[27]

Thus the original Plan Hoog Catharijne of 1962 consisted of an all-out redesign of Utrecht's railway station district. The plan covered the whole area in the immediate vicinity of Central Station. The main argument for choosing this location was tied to its central role as a national railway junction, while it served as a transportation crossroads at a local and regional level as well. Furthermore, it was argued that this area could be elegantly connected to the pedestrian routes of the old inner city of Utrecht. A radical reconstruction of the entire Central Station area would also provide ample opportunities for solving the city's traffic and parking problems.[28] For this reason, the infrastructure of the area had to be transformed drastically. It would include office buildings, apartment buildings, parking lots, a new railway station, a concert hall, a bus station, and the largest indoor shopping mall in the Netherlands. The design of Hoog Catharijne was based on strict separation of traffic flows. The connection between the station area and the old part of the inner city was to be established by the construction of separate pedestrian passages. (See figure 2.3.) Because the total space for the various forms of mobility—pedestrians, cyclists, cars, and buses—had to be augmented, various functions were to be put on separate levels.[29] The pedestrian level, comprising the indoor mall and Central Station, would be 5.5 meters above street level.[30] The entrances of shops and offices would be situated at this level, while the street level was reserved for parking, service roads, the bus station, showrooms, and so on.[31] The elevated level not only provided a solution for the barriers posed by the Catharijne Canal and the railway tracks, it also created a shorter pedestrian connection between the railway station and the old part of the inner city. This functionalist design allowed people to shop without being disturbed by car traffic or other nuisances.[32]

The first plans included broad open-air passages for pedestrians rather than indoor ones. The initial idea was to construct open pedestrian squares at a raised level. (See figure 1.1.) However, this plan was abandoned for safety reasons after wind-tunnel tests showed it to be dangerous. Because of its openness and the unpredictability of wind gusts in high-rise areas, people might be "blown away."[33]

To be sure, commercial motives also in part determined the ultimate plan: Hoog Catharijne's design more or less compelled pedestrians to walk through passages lined by shops because this was the easiest and most comfortable route between the railway station and the city center. This arrangement was expected to turn train passengers into potential customers. The estimated number of pedestrians in Hoog Catharijne was based on the expectation that people would choose the shortest route, which meant that sections of the mall were likely to become as busy as the most crowded shopping street of Amsterdam. As a result, it would be very attractive to retailers.[34]

In this way, preferences for functionalist planning and traffic regulation and calculations about risks and commercial potential were built into Hoog Catharijne's design. The plan was accepted almost unanimously by the City Council, and in February 1964 the city signed a contract with development company Empeo.[35] A new company, in which the city did not participate, was founded for the realization and exploitation of the plan. The company was called Hoog Catharijne Ltd. The contract basically states that the city leases the area to Hoog Catharijne Ltd. until the year 2070.

A few obstacles had to be dealt with, however, before and during the building process. As a consequence of situating the new plan in the station area, the existing houses near Central Station had to be demolished. These demolition plans were very much in line with the trend to replace functions and properties that were considered unprofitable, such as houses with low rental values or small shops and cafés, with more profitable venues.[36] Yet evidently the residents whose houses were going to be sacrificed were not all happy about it, and they protested against the plan. Another problem involved the nearby convention center of the Royal Dutch Industries Fair,[37] which had to be moved. The Industries Fair is a huge complex of buildings that can be rented for exhibitions, conferences, fairs, leisure activities, and so on. Although the Industries Fair was already coping with a lack of space in 1963, it did not want to move to the west side of Central Station. Empeo had great difficulty developing a plan in which the buildings of the Industries Fair could remain in place. Finally, in 1968, the Industries Fair decided to move to a nearby location.

The debates about converting some canals to roads went on for nearly 10 years. In 1964, Utrecht's medieval canals provisionally acquired the status of national monument.[38] As a consequence, the canals could not be tampered with without permission from the Dutch government. Nevertheless, in May 1964, the City Board proposed to fill in the Catharijne Canal, a plan that readily passed the City Council. In order to win the national government's consent, a working group consisting of representatives from the ministries, the region, and the city was founded. It concluded that filling in the Catharijne Canal was necessary for the realization of the Plan Hoog Catharijne, suggesting that Feuchtinger's plan to convert some of the canals into roads had become so much built into the subsequent plans that it was no longer sensible to avoid it. Finally a compromise was reached and Minister Marga Klompé granted permission to fill in part of the Catharijne Canal in January 1968.[39]

In 1969 the first new office buildings, parking garages, and one of the raised pedestrian passages of Hoog Catharijne were completed. Although the building activities would continue until the late 1980s, the official opening of Hoog Catharijne took place on September 23, 1973. (For an overview of the area, see figure 2.2.)

Hoog Catharijne's planning design thus became gradually embedded in Utrecht's urban structures. Many have asserted that Hoog Catharijne gave a boost to the local economy, but in the 1980s some negative consequences came to the fore. Drug addicts, homeless people, and psychiatric patients populated the indoor mall in ever-larger numbers, and the dark service alleys and ground-level parking areas became meeting places for drug dealers and addicts.

In 1982, the Hoog Catharijne shopping mall was bought by Algemeen Burgerlijk Pensioenfonds (ABP), a large Dutch pension fund. ABP intended to revitalize the property after many stores had seen their profits decline and some of them went bankrupt; the area's other large property owners, the railway company NS and the Industries Fair, were also making plans for improvements. In 1987, the newspapers reported that the city initiated the development of a new plan for

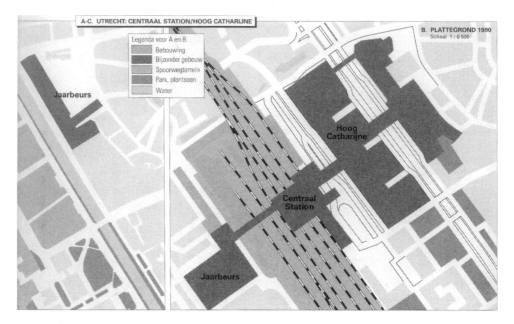

Figure 2.2
Overview of Hoog Catharijne area. Adapted from *De Grote Bosatlas*, 1998. © *De Grote Bosatlas*.

the Central Station/Hoog Catharijne district. It wanted to design a completely new plan for the area in cooperation with NS, the Industries Fair, and ABP.[40] This public-private partnership was in charge of the Utrecht City Project,[41] and in June 1988 the partners signed a declaration of intent to develop the new plan.[42]

The plans for the redevelopment of Hoog Catharijne and the railway station area can be situated in a trend of redevelopments of railway station areas throughout Europe. In the late 1980s and the 1990s several European cities (e.g. Stockholm, Basel, Zurich, London, and Lille) developed ambitious plans for new stations. The wish to build a new network for high-speed trains was also a major motivation to initiate such projects. In their book *Cities on Rails*, the urban researchers Luca Bertolini and Tejo Spit analyzed the redevelopment of station areas in the above-mentioned European cities, including Utrecht. What, they asked, makes the redevelopment of railway station areas such a daunting task?

As one of their examples, Bertolini and Spit investigated the reconstruction of the King's Cross station in London. King's Cross/St. Pancras is one of the most intensively used transport interchanges in the United Kingdom. The ambition of the redevelopment plans was to turn King's Cross into a large international station. One of the reasons why it was so difficult to make this project successful was that "power was unevenly distributed" in the partnership that was in charge of the project. A more successful project, according to Bertolini and Spit, was the development of a new "Travel Center" in Stockholm—a project consisting of a huge expansion of the railway station and the building of a new bus station and offices. Bertolini and Spit claim that the main reason for this project's success was the "professionality" of the partnership that initiated this project: a combination of the City Board and private developers.[43]

THE GROUND-FLOOR MODEL VERSUS THE RAISED-LEVEL MODEL

From the outset, discussions about the new plan in Utrecht in the late 1980s were dominated by a distinction between two models: the "ground-floor" model and the "raised-level" model. The ground-floor model was advocated by the city, whereas the raised-level structure was strongly preferred by ABP. What were the basic views advanced by the two parties? The Director of the Planning Office of the city of Utrecht formulated the central problem as follows: "The basic issue of the two models presented here involves the question whether the rigid separation between having pedestrians on an elevated level and the other flows of traffic on the ground floor level, that was favored in the 1960s, can be reversed by linking up pedestrian traffic with the ground floor level."[44] The "ground-floor model" emphasized the accessibility of public buildings—including Central Station—at the ground-floor level.[45] Public safety also played a prominent role. Especially at night, the dark and deserted parking spaces and service alleys on the ground floor had attracted criminal activity; it was argued that an upgrading of this level, in part by encouraging the presence of more pedestrians, would improve public safety at this level. Therefore, this model emphasized the significance of adding alternative walking routes. It should still be possible to walk through Hoog Catharijne via the raised level, but alternative routes had to be added that contributed to a better integration of the ground-floor level in the overall structure.

The ground-floor model followed ideas on urban design that gained prominence in the late 1960s. During that time, authors who wrote about planning began to highlight the "social city"—the city as meeting place.[46] The American Jane Jacobs is often portrayed as the most important proponent of this tradition.[47] In her book *The Death and Life of Great American Cities* (1961), Jacobs harshly criticized contemporary American urban planning, especially condemning the megalomania and totalitarianism of the large-scale urban interventions proposed by Le Corbusier and Robert Moses. Rather than large-scale, monotonous, functionalist planning, she emphasized the importance of small-sized building, diversity, (social) safety, and concentration. In her book, Jacobs describes the city from the perspective of the neighborhood resident, the pedestrian, to whom the street and the sidewalk are indispensable symbols of the city as a meeting place. She argued that the street should be re-established as a major urban public space.[48] Jacobs's views found resonance both in Europe (including the Netherlands[49]) and in the United States. In line with the activist spirit of the late 1960s and the early 1970s, local activist groups and neighborhood committees were established in response to undemocratic decision processes and large-scale urban interventions. Proponents of the ground-floor model were in part inspired by this new mode of thinking about city life.

The "raised-level" model basically held on to the existing philosophy of separating flows of traffic. The pedestrian passage that cuts through Hoog Catharijne should continue to be the main route, proponents argued, but it had to be improved. In their view, the new design should enhance the raised-level structure by expanding it.[50] The creation of new pedestrian routes on the ground-floor level, as the other model proposed, would decrease the amount of pedestrians at the raised level, and this was seen as a negative factor in public safety at the raised level.[51]

The two models were presented in a report titled The Utrecht City Project: Perspectives for the Future.[52] Although it emphasized that the two models represented two extremes, the ensuing debates made it seem as if a choice had to be made between them, and gradually parties were formed that defended the two models.[53]

The city expressed a preference for the ground-floor model, arguing that the main problem was poor public safety and that this was attributable to flaws in Hoog Catharijne's original design. Because of the elevated pedestrian passages, the ground-floor level—dominated by motorized traffic—had little value as a public space. The city representatives stressed the significance of the street as a public space and the role of the city as a meeting place.[54] The quality of the ground-floor level as a public space, they asserted, could be improved by introducing additional social and commercial functions at that level, such as shops and apartments.

According to ABP, the increasing number of drug addicts and homeless people in Hoog Catherijne caused great inconvenience, fostered a negative image of the shopping mall, and therefore threatened its profitability. ABP wanted to control this problem by installing cameras, refurbishing the mall's interior design,[55] and concentrating pedestrians on the raised level to maximize social control. Although ABP was interested in improving the safety of the area, from a business point of view it argued against an effort to integrate the ground-floor level. An exclusive focus on the ground-floor level presented obvious risks for the functioning of the mall, and thus for the profitability of its shops, all of which had their entrances at the elevated level[56]; therefore, this level ought to be preserved as the main route. Radical changes in Hoog Catharijne's basic design were not called for, because the mall—despite some of its obvious problems—functioned quite well as a comfortable connection between the inner city and Central Station. A former ABP manager suggested that the separation of traffic flows was "one of the greatest achievements" of the original design.[57]

The main problem for NS was that in the 1970s Hoog Catherijne had swallowed Central Station, which had thus become invisible and lost its identity. The NS's main concern in light of the City Project was to turn Central Station into a visible urban entity once again.[58] This wish complied with the trend of increasing importance attached to the architecture of railway stations in the Netherlands since the 1980s. Station architecture in the 1970s was characterized by austere, standard buildings, whereas since the early 1980s NS had tried to make its stations more "clear, convincing, recognizable."[59] In this respect, NS saw Hoog Catharijne as a major obstacle in re-establishing a direct link to the old part of the inner city. In 1989, NS was not convinced by either of the two models. The recognizability of the railway station remained a problem in both models, NS felt, but holding on to and expanding the raised-floor model would be the worst solution.[60]

The Industries Fair had relocated to the west side of Central Station and Hoog Catharijne in the late 1960s. Like NS, it considered Hoog Catharijne a barrier that made it difficult to visitors to the Fair to reach the inner city—with its many cafés and restaurants—by foot. Thus, the main interest of the Industries Fair was to establish comfortable connections between the inner city, Central Station, and its own premises.[61] The raised-floor model comprised a new, extended pedestrian passage that significantly improved the connections between the Industries Fair, Hoog Catharijne, and the old part of the inner city. Not surprisingly, the management of the Industries Fair favored this model.[62]

Clearly the Perspectives report did not lead to a consensus among those who had major stakes in it. On the contrary, it pitted the various parties against one another, with ABP and the Industries Fair opting for a plan based on the elevated model and the city and NS strongly favoring the ground-floor model.

THE EMERGENCE OF TWO DOMINANT TECHNOLOGICAL FRAMES

It is clear that at this stage in the Utrecht City Project it was difficult to achieve consensus and to undertake real unbuilding activities. How can the obduracy of Hoog Catharijne in this stage of the project be explained? It is clear that the city and ABP were trapped in a discussion about two opposing models. I will argue that the first model of obduracy, the category of "frames," offers some useful conceptualizations. The concept of technological frames seems the most appropriate, because, in contrast to the other conceptions in this category, technological frames are also applicable to other social groups than engineers, such as citizens. Moreover, this concept is sufficiently broad to encompass the heterogeneity of elements that are involved in urban socio-technical redesign.

Bijker proposes to distinguish three "configurations," based on the different roles of technological frames in the interactions between actors and the processes of socio-technical change and stabilization that are expected to take place.[63]

In the first configuration, there is no clearly identifiable dominant technological frame, nor is there a single dominant social group. In the absence of overriding vested interests in the interaction process, radical alternatives are likely to emerge. A redefinition of the problem usually functions as the closure mechanism in such a configuration. This means that social groups try to persuade other social groups of their view by redefining the problem in such a way that it becomes attractive for this group to support the socio-technical ensemble.

In the second configuration, one technological frame dominates the interactions of various actors. In such cases, Bijker finds it useful to make a distinction between actors with high inclusion and those with low inclusion in this technological frame. The interactions of actors with high inclusion are very much structured by the technological frame involved, which usually results in conservative alternatives on their part. Actors with low inclusion are less "guided" by the technological frame, and, as a result, are more likely to come up with more radical variations.

The third configuration is characterized by the presence of two or more equally dominant technological frames, as well as two or more dominant social groups. In this configuration, arguments that play a crucial role in one frame are mostly seen as irrelevant in the other frames. In such cases, rhetoric plays an important role in the stabilization process. Frequently, the result is a situation in which "no one wins a total victory."[64] An "amalgamation of vested interests" results in moderate modifications at best, because, after all, the proposed innovations have to be acceptable to all groups and have to fit all the dominant technological frames.[65]

When we look at the situation in Utrecht we can see that during the discussions between the city and ABP about these models two dominant technological frames emerged: one based on the ground-floor philosophy, advocated by the city, and another taking the raised-level structure as its point of departure, strongly preferred by ABP. (For a summary of the elements of these two technological frames, see table 2.1.) This situation of two dominant technological frames corresponds best with Bijker's third configuration. In the following section I will analyze how, exactly, the ground-floor frame and the raised-level frame structured the interactions between the actors involved in the planning process and how that resulted in a situation in which change in the established situation became increasingly difficult to accomplish. Arguments that played a role in the ground-floor frame were not considered to be important in the raised-level frame. The fact that actors were thinking and acting so much in terms of their own technological frame made it well nigh impossible to bring about any change at all, thus contributing to the obduracy of Hoog Catharijne.

1989–1995: Struggling with Fixed Urban Structures

For a long time after the Perspectives report was published, the redesign options were limited to versions of either the ground-floor model or the raised-floor model. (For an overview of the main UCP plans published between 1989 and

Table 2.1 The ground-floor frame versus the raised-level frame.

	Ground-floor frame (city)	Raised-level frame (ABP)
Goals	Improving quality and livability of ground floor	Strengthening and improving (profitability of) raised level
Definition of problems	Poor architectural quality and limited public safety of street level	Decreased economic profitability of raised level, poor image of raised level
Problem-solving strategies and designs	Creating new pedestrian axes on ground floor, demolishing parts of raised level, adding functions like stores and apartments to ground floor, improving morphological cohesion between Hoog Catharijne and inner city	Creating a circular route at the raised level, broadening raised passages and adding shops along them
Current theories and models	"Compact city" model (governmental policy guideline)	Business economics, studies of consumer behavior
Design principles	Mixing traffic types, street as public space (accessible to pedestrians and cyclists)	Preserving vertical segregation of traffic types
Requirements to be met by problem solutions	Improving liveability, meeting governmental guidelines and policy aims	Profits may not decrease

Table 2.2 Official UCP plans published between 1989 and 1997.

Perspectives report (February 1989)

Interim report (November 1989)

Project plan (June 1991)

UCP master plan (May 1993)

Spatial-Functional concept (June 1995)

Preliminary town plan (February 1997)

Definitive town plan (October 1997)

1997, see table 2.2.[66]) In 1989, it was hoped that by investigating the traffic aspects, a choice could be made between these two models.[67] This study resulted in the "interim report."[68] Apart from an investigation of the traffic aspects, the aim of this report was to adapt the plans to incorporate demands and regulations put forward by the national government. In the early 1990s, governmental policies on zoning in urban areas encouraged the concentration of employment near junctions of public transportation. Moreover, these policies recommended the "compact city model," which implied that urbanization should be concentrated in areas situated relatively close to city centers. Use of public transportation had to be stimulated and car usage had to be decreased. Governmental policies formed an important element in the ground-floor frame, and the city wanted the UCP plans to be in line with these policies, partly because this would improve its chance to obtain substantial governmental subsidies to support the project. Statistics and prognoses played a prominent role in the city's effort to justify the new plans that were based on variants of the ground-floor model. The city's strategy to link it with governmental policies and subsidies proved to be a powerful tool in stabilizing it. The interim report presented statistics to support the potential strength of the UCP area as a top location for service industries: many companies expressed a preference for the city center of Utrecht as a business location. On this basis it was predicted that the number of jobs in Utrecht's downtown district would go up. In view of the governmental policies aimed at stimulating the use of public transport and decreasing car usage, an increase of the passenger flow in Central Station was projected. On the assumption that Hoog Catharijne would not be able to absorb this increase, it would be necessary to add alternative pedestrian routes; the interim report proposed such a new pedestrian route at the street level between the inner city and Central Station. This proposal was explicitly based on "a variant of the ground-floor model."[69]

After the interim report was discussed in public hearings and in municipal commissions, it was formally accepted on December 13, 1989 as a guideline for the plans that were to follow.[70] This adoption played a crucial role in stabilizing the ground-floor model as the single basis for the future UCP plans. Even though the raised-floor model was thus more or less abandoned, ABP still had severe doubts about this choice.[71]

In 1990, the city took the initiative to establish the "UCP Atelier," consisting of a group of planners. It developed a proposal in which the main pedestrian axis on the raised level that cuts through the mall, the Radboud Passage, would have an equivalent on the ground-floor or street level. (See figure 2.3.) ABP, the owner of the mall, reacted furiously to the plan. Because of the unrest it caused among the partners, the planning group—after a turbulent period of six weeks—was disbanded.[72] In ABP's view, the plan entailed the demolition of a crucial part of Hoog Catharijne.[73] Serving as the main pedestrian artery from the old part of the inner city to the railway station, the Radboud Passage was an essential part of Hoog Catharijne, according to ABP. Because most of the mall's shops were located along this passage, it generated major profits for its owner.

As was noted above, ABP wanted to expand and remodel Hoog Catherijne rather than demolish sections of it. Already in 1987, ABP and the management office of Hoog Catharijne were making concrete plans to "revitalize" the mall. The outdated interior of Hoog Catharijne had to be transformed into a more attractive design with more daylight, bright colors, and newly designed furniture. One of the underlying goals was to make the mall less attractive to homeless people and drug addicts.[74] The director of the management office, a man named Macrander, proposed that a second main pedestrian route through Hoog Catharijne be created by widening the Gilden Passage.[75] According to ABP manager Ed Bolt, it was important for a shopping center to function well in its totality. The Radboud Passage was always crowded, but the rental spaces in the "Gilden quarter" section were not doing well at all, which meant declining profits for the company. Therefore, the Gilden Passage played a crucial role in ABP's effort to construct a pedestrian circuit at the elevated level. To establish a more even spread of the visitors throughout the complex, the Gilden Passage had to be enhanced by lining it with shops.[76]

Instead of viewing the Gilden Passage as a potentially vital element in the functioning of Hoog Catharijne, the city stressed that this section of the complex was not operating well. City Board member Ger Mik dismissed the

Figure 2.3
Radboud passage and Catharijnebaan. Photograph by A. Hommels. © A. Hommels.

proposal to improve it as "not right" and "not sensible."[77] Because the city considered the Gilden Passage an ineffective connection between the old part of the downtown district and Hoog Catharijne, it even proposed to tear it down and replace it with a street-level passage. In contrast, in ABP's technological frame the overall enhancement of the raised level was a major goal. ABP reasoned that this level's commercial potential would benefit greatly from a circular pedestrian route, a widening of its passages, and the introduction of shops along the Gilden Passage. Because of the company's high inclusion in this technological frame, it could only generate alternatives within the parameters of the existing design. Thus the obduracy of the Radboud Passage emerged as the major factor in the negotiations.

In December 1990, the UCP was designated a potential "key project" by the national government. This status implied that the project, because of its national significance, might be eligible for substantial governmental subsidies, which, in turn, was a strong impetus for incorporating governmental policies and rules in order to attain "key project" status. This led to the "Project Plan,"[78] which was designed to incorporate these governmental guidelines.

At this stage, the ideal of the compact city, combined with NS projections of a doubling of the number of public transportation passengers and data that signaled an increase of the overall population, caused the city to believe that the number of passengers at Utrecht Central Station was likely to go up even more than was assumed initially.[79] To meet the anticipated demand, NS was planning to build a second railway station near Central Station. Obviously, this would have major consequences for the pedestrian routes connecting both stations and the inner city. ABP, however, disagreed with the data put forward by the city and NS; it projected that pedestrian activity would increase only 6 percent.[80] ABP acknowledged that Hoog Catharijne would be unable to accommodate twice as many passengers. Since ABP's technological frame was mainly driven by commercial concerns, the company stressed that solutions should consider not only the planning aspects but also the business economics.[81]

The city and NS nevertheless succeeded in integrating their "data" into the new plan, which proposed four pedestrian axes on the ground floor instead of one central axis at the raised level. Moreover, it called for a new ground-level axis that would cut through the existing Hoog Catharijne buildings at Vredenburg Square. Clearly, it was hardly a plus in the city's frame to maintain all the existing Hoog Catharijne buildings. The street leading to the Station Square had to be improved by adding street-level shops, while at the same level new entrances to Central Station were needed. Moreover, the Project Plan mentioned the idea of returning water to the Catharijne Canal, which had been paved over in the 1960s.[82] The city considered this element in the plan of crucial importance. Ger Mik, the City Board member in charge of the UCP between 1990 and 1994, saw the recovery of the canal as proof of the city's commitment to recovering urban public space.[83]

Although the city and ABP still disagreed, the city decided to submit the Project Plan as the basis for governmental subsidies.[84] ABP objected to the plan for commercial reasons: it would leave a deficit of 300 million guilders. Moreover, the company feared profit losses: the city's plans would cause a decline of potential customers, because with street-level pedestrian routes people would no longer be obliged to enter and cross Hoog Catharijne.

Despite the ongoing disagreement with ABP during the next stage of the process, the city held on to the idea of creating street-level pedestrian axes. In an effort to prevent this idea from becoming a reality, ABP brought into play official agreements signed by the city in the 1960s. If the new central street-level axes were put in place, official agreements dating from the 1960s would be broken.[85] This legal threat did not keep a new central axis on the ground floor out of the UCP Master Plan that was published in May 1993. The UCP Master Plan revived the grid[86] as a central concept. A grid is an orthogonal screen of public streets that, in principle, are equally important. The grid structure of the area had to be improved by adding a new axis at the ground-floor level. To create this axis, a 22-meter-wide passage had to be cut through the existing buildings of Hoog Catharijne. The new axis would be the shortest route from the old town center to the entrance of the railway station. This axis was presented as a major feature of the ground-floor model. For ABP, however, the new axis was unacceptable if it would facilitate a shorter and more comfortable connection

Figure 2.4
Artist's impression of recovery of Catharijne canal proposed in UCP Master Plan. Source: UCP Masterplan (1993).

between Central Station and the old city center.[87] Moreover, Ed Bolt opposed the idea that the plan called for a 22-meter passage, whereas the Radboud Passage, the main raised pedestrian passage through Hoog Catharijne, was only 9 meters wide.[88]

ABP did not agree with the Master Plan and did not sign it. The company's main objections concerned the economic feasibility, the new central axis, and the Gilden Passage.[89] Moreover, NS refused to sign the financial paragraph. In the meantime, however, the declaration of intent with NS, the Industries Fair, and ABP had expired, so the city had an opportunity to continue the planning process with new partners. In May 1994, the Development Company Ltd. was founded, a collaborative effort of the city and a group of project developers.[90] This company aimed to develop a feasible plan based on the Master Plan. Owning 51 percent of the shares, the city had a very strong position in the Development Company; it could force specific decisions in this way—and it did.

This collaborative effort resulted in a new plan, called the Spatial-Functional Concept.[91] This plan was very much in line with the city's ground-floor model: "With its decision to upgrade the public environment at ground level, the council chose to add a fully fledged piece of city at ground floor level to the old inner city, and to declare large areas of the public space at street level to be the domain of the pedestrian and the cyclist."[92] Moreover, it was emphasized that the enhancement and re-establishment of functions at the street level would improve the "morphological cohesion" of the modern Hoog Catharijne complex and the old inner city.[93]

The plan again proposed a street-level axis between Vredenburg and Station Square. It thus provided two possibilities to reach the railway station: via the raised level of Hoog Catharijne and via street level. In the plan, the new ground-floor axis directly led to a completely new railway station. The new station's central hall was connected to Hoog Catharijne only by a pedestrian passage.[94] The Gilden quarter area, connected to Vredenburg by a raised passage, would have a new street-level entrance. Furthermore, the ground-floor domain was to be improved by the introduction of shops and cafés.

After the publication of the Spatial-Functional Concept, it turned out that the former partners (ABP, Industries Fair, and NS) strongly objected to the plan. Not only were they upset about not being consulted during its development, they also felt that the plan violated their rights. They were all but prepared to give the city and the project developers permission to rebuild their property on the basis of the new plan. Moreover, after its publication it became clear that the project developers were not entirely satisfied with the plan either. In the absence of a fair decision process and because of a lack of confidence between the city and the project developers, their collaboration was discontinued.[95]

During the period between the Perspectives report (1989) and the Spatial-Functional Concept (1995), the city succeeded in pushing decisions that were increasingly irreversible and that profoundly influenced the later plans. For one thing, its preference for the ground-floor model became incorporated in all subsequent plans. Statistics, projections of increasing pedestrian flows, governmental policies, and City Board decisions were deployed to promote plans based on the ground-floor model. Although ABP tried to destabilize the plans with reference to previously signed contracts and cost-benefit analyses, the officially published plans show that the ground-floor model increasingly functioned as the major planning guideline.

Closed-In Obduracy

As was suggested above, two technological frames structured the controversy in which the city, ABP, NS, and the Industries Fair were the main players. Between 1989 and 1995, the two groups of actors had great difficulty developing alternative designs for the Hoog Catharijne area that would fit either the ground-floor model or the raised-level model. Thus, I would argue that these actors' ways of thinking and interacting can be described in terms of closed-in obduracy.[96] This phrase refers to actors with high inclusion in a particular technological frame. Because of their high inclusion, these actors think and interact very much in terms of this technological frame, and they are unable to conceive of alternatives outside this frame. This results in a limiting of the available redesign options to mere variations on the existing schemes. "Closed-out obduracy" refers to actors with low inclusion in a technological frame. These actors see little or no opportunity for variation within the frame, and so they are prone to adopt a "take it or leave it" approach; this generally triggers the formulation of radically new alternatives.

My analysis so far demonstrates that Hoog Catharijne's obduracy was closely tied to closed-in obduracy. ABP and the city took stances that suggested their high inclusion in a particular technological frame. These two actors could only think in terms of the raised-level frame and the ground-floor frame, respectively. They only saw alternative designs or tried to stabilize designs that fitted their frame. Arguments that played a major role in one frame were not considered relevant in the other. This limited the available redesign options to versions of the raised-floor model (nearly every one of which involved an effort to raise the profitability and effectiveness of Hoog Catharijne's existing raised-level configuration) and versions of the ground-floor model (nearly every one of which was motivated by the desire to better integrate the ground-floor-level functions into the overall structure so as to improve public safety and the overall quality and livability of urban space). The closed-in obduracy that resulted from ABP's and the city's high inclusion in their respective frames had far-reaching consequences for the planning and decision processes: the rigidity of these opposing frames caused a deadlock in the planning process.

CITIZENS' PLANS FOR THE REBUILDING OF HOOG CATHARIJNE

Despite the residents' opposition to Hoog Catharijne in the 1970s, the complex gradually became accepted as a fact of life. Even in a 1997 newspaper article, someone involved in the colossal complex's initial design still commented that it has such presence that "it cannot be ignored by anyone and that we are all forced to take it into consideration."[97] Over the years, city residents began to take Hoog Catharijne for granted in its existing form without attempting to change its character. However, in the course of the 1990s some groups of residents became actively involved in the plans to change Hoog Catharijne. In 1990, a group of concerned residents, who wanted to influence the planning and decision processes involving the Utrecht City Project, founded the Bewonersoverleg City Project (Residents Committee City Project).[98] It argued that the UCP should primarily pursue the re-establishment of public safety and the improvement of the livability of the area around Hoog Catharijne. Although the Bewonersoverleg City Project (BOCP) never explicitly expressed a preference for either the ground-floor model or the raised-floor model, it favored the former. Among other things, it argued that the existing separation of traffic flows should be given up in part and replaced by an integrated flow of pedestrians, cyclists and public transportation.[99] The BOCP argued that the planners should pay extra attention to the architecture at the ground floor, that houses should be planned at the street level, that safe walking routes from the city center to the railway station should be created at the street level,[100] that pedestrians and cyclists should have absolute priority, and that automobile traffic ought to be limited drastically in the area.[101] Although these ideas seem well in line with those of the city, the residents had different priorities. The BOCP argued that the UCP's goals had shifted as a result of governmental policies in the field of urban planning.[102] In contrast to the city, the BOCP saw a tension between the goals of governmental policies and the need to improve the livability and public safety of the city. It stressed the importance of bringing back a sense of "human scale" to the area. This meant that large-scale building had to be avoided and that the preconditions for public safety and urban livability had to be specified. Subsequently, it could be determined which functions would be necessary to finance this.[103] Leo Lambo, former coordinator and secretary of the BOCP, suggested that the BOCP was never against the UCP: "We have always said: 'The UCP was necessary, because something had to be done to improve the area. But the surrounding neighborhoods should not be negatively affected.'"[104]

Some residents, who believed that minor adjustments of the existing Hoog Catharijne design would not solve the area's public-safety problems, proposed demolishing the complex. This would create, according to one resident, building space for a new plan that would be more "humane." The same resident argued that it would be impossible to solve the public-safety problems if the plans essentially implied a continuation of the present structure.[105]

A detailed plan for the demolition of Hoog Catharijne was put forward by the Utrecht architect Clemens Koemans. (See figure 2.5.) Koemans was very critical of the UCP plans and entirely displeased with the existing architecture of Hoog Catharijne. In 1993, he participated in meetings at the Utrecht Architects Café in which experts elucidated the UCP plans as a way to involve the general public. Expressing his criticism at one meeting, Koemans received support from Leo Lambo of the BOCP. Later he contacted Lambo and communicated his idea of making an alternative plan based on his criticisms of the UCP. Koemans:

> Lambo told me that I could get all the information he had. He played a pivotal role in the alternative scene. I have always regarded them as a client in a way. But I have done this voluntarily, without being paid for it. . . . The BOCP has never fully accepted my plan. They sympathized with it, thought it a nice alternative, but I introduced the plan as my own.[106]

Koemans proposed his plan for a city park on the location of Hoog Catharijne. He argued for a "humanist" planning perspective. People were compelled to pass through the mall. Koemans felt that they should have the option to choose other routes. Because it would be impossible to implement a new shopping concept in the present structure of Hoog Catharijne, he called for complete demolition of the mall.[107] The plan was rejected on the grounds that "to ascertain that a demolition of the Hoog Catharijne complex is unfeasible" hardly needed elucidation.[108]

Closed-Out Obduracy

As my analysis makes clear, the specific design proposals made by another actor involved in the negotiations involving Hoog Catharijne's redesign, the concerned residents of Utrecht, did not fit either of the two dominant models

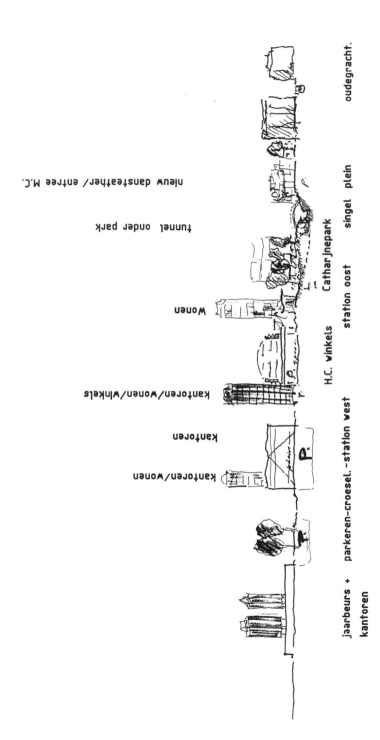

Figure 2.5
The Koemans plan. © Clemens Koemans.

that were supported by the other actors. The role of Utrecht's citizens and their planning proposals can be understood when the low inclusion of this group in these two dominant technological frames is considered. Free from rigid frames, some city residents made radical redesign proposals for Hoog Catharijne. This did not imply, however, that this freedom also gave them more possibilities to influence Hoog Catharijne's future, or that they could bend things to their will. I argue that what Utrecht's residents were faced with can be understood in terms of Bijker's analysis of closed-out obduracy.

Hoog Catharijne has such an uncompromising presence to some residents and some users that these actors can only conceptualize it in terms of a "take it or leave it" choice. Actors with low inclusion typically either propose radical alternative technological designs or decide to accept a technology as it is. Both strategies are visible in the history of Hoog Catharijne. Instead of trying to adapt the existing design of Hoog Catharijne, some residents completely dismissed it. Whereas the city and ABP tried to think of alternatives within their technological frames, these residents—being closed out of the existing frames—saw no alternative designs within these frames. In their view, Hoog Catharijne should be either demolished or accepted.

My argument in this chapter demonstrates how technological frames influenced the plans for redesigning Hoog Catharijne and how the actors were caught up in closed-in and closed-out obduracy. But was change possible outside these frames? Is it possible to transform or abandon dominant technological frames? If so, in what ways? In other words, how can these types of obduracy be overcome?

OVERCOMING OBDURACY

Despite the city's efforts to promote a serious redesign of Hoog Catherijne's ground-floor level as a significant contribution to the overall recovery of Utrecht's downtown district, the Utrecht City Project became increasingly associated with failure. In 1995, to mark a new start, the name of the project was changed to Utrecht Centrum Project.[109] After the failure of the Development

Company in that same year, the city decided to re-open discussions with its former partners. It was fully aware of the importance of renewed cooperation, since an investigation of the failure of the Development Company by a consultant had shown that close collaboration between the partners would be indispensable for a successful plan, which, after all, depended on support from various actors.[110] The city and its partners were convinced that their negotiations had reached a "now or never" stage; they realized that this was their last chance.[111] A new official agreement comprising new financial arrangements was signed by the four partners. It was decided to divide the UCP area into four parts, one for each partner. Each partner would take the financial risks for its area. Infrastructure and public space would be financed by subsidies from the national government and by profits from exploitation of the whole area.[112]

In 1995, the mayor of Utrecht, Ivo Opstelten, asked Riek Bakker, a planner and a former director of development for the city of Rotterdam, to become Utrecht's new supervisor of planning. Bakker was particularly interested in the *process* of planning and its strategic aspects:

> Bringing together all ... vested interests takes a lot of time, energy and, in particular, much patience. One of the reasons is that interests will often contradict each other. Of course, the town planner has to avoid becoming a (political) "party" in such a process. Everyday practice suggests that political and diplomatic insight is desperately needed. The biggest chance of success, however, is based on the approach taken: from the outset the town planner looks for a solution to the issue as a whole, and this includes taking into account all the various interests.[113]

Her initial task was to further develop a plan on the basis of the Spatial-Functional Concept.[114] Bakker, however, was not entirely convinced of the quality and feasibility of this plan. She wanted to start from scratch by analyzing the concrete problems in the Hoog Catharijne area. In this process, she devoted particular attention to the vested interests in the area and to the views and preferences of

---

the partners.[115] By taking into account the history of the two technological frames and by confronting the partners with their dogmas, she developed a new plan that got the support of Utrecht's City Council and the partners.

A preliminary plan was published in February 1997.[116] Although this plan was presented as the "logical" next step after the Spatial-Functional Concept, it was an altogether new plan based on different assumptions.[117] The key feature of the plan was a large inclined square, a bridge between the ground floor and the raised level. (See figure 2.6.) To allow for the construction of this square, the entire Gilden quarter area had to be demolished, and a part of the remaining Hoog Catharijne complex had to be split in two. The square was to function as a connection between the street level and the railway station. The planner Evelien Brandes pointed out that Central Station, instead of being an integral part of the shopping center, ought to be an independent building, and that the inclined Station Square should be an open-air public space.[118] Having arrived at Central Station, passengers should be able to enter the square at the street level or proceed through the mall. The station entrance, however, was still situated at the raised level.[119] The idea to return water to the Catharijne Canal was preserved from earlier UCP proposals. The Radboud Passage was to be widened, and Hoog Catharijne was to be rearranged in the style of American shopping malls.

Bakker's plan seemed to harmonize well with many of the partners' points of view, but the plan also included elements that strongly conflicted with positions that had been articulated earlier in the UCP process. How did Bakker manage to achieve a reconciliation of such diverging ways of thinking? What was so remarkable about her plans and strategies that the partners were willing to forsake their firmly established positions?

Bakker's strategic talents and her experience with the procedural aspects of planning contributed significantly to her success in the UCP. She was praised as a "real personality," for her "outspoken language," and for her ability to say the right thing at the right moment. "She does a lot of 'massaging' behind the scenes," said Bart Lambooy, Bakker's advisor in the UCP. Moreover, "She functions in a man's world. And one way or another, she knows exactly how to handle that world, twisting all those men round her finger."[120] Bakker herself

Figure 2.6
Artist's impression of inclined square (1997). Source: Projectbureau UCP Utrecht.

characterized her role in the UCP as that of "organizer, communicator, *postiljon d'amour*, binder, trustee." [121]

Bakker also took advantage of the many changes that had occurred in the corporate philosophies of ABP, NS, and the Industries Fair. Their views, issues, and goals had gradually become defined differently. ABP had reorganized its investment funds and had founded WBN (Winkel Beleggingen Nederland), a shopping center investment fund. WBN had become a partner in the project. NS had been privatized. New ideas had emerged about organizing public transport and integrating different public transport systems (bus, tram, train) at one location, which led to a new "mainport" concept for Utrecht Central Station. The Industries Fair had developed a new business philosophy, and this had major consequences for the Fair's goals in the UCP.

At this stage, several crucial redefinitions resulted in an increased malleability of Hoog Catharijne. Instead of choosing the ground-floor model as the basis for the plans, the partners agreed that the raised level should be the point of departure. [122] The project manager responsible for the UCP, Ad Smits, said: "The dogmatic discussion about shopping at street level, street-level access to the railway station, and the competition this would imply for Hoog Catharijne, which is situated 6 meters above street level, was solved because the parties agreed from the outset of this new phase: the raised level is the point of departure. . . . There has been a profound change in the city's way of thinking." [123] Herman Kernkamp, a City Board member responsible for the UCP, argued that Bakker brought the debate to an end by making clear at the beginning that it was a great misunderstanding to speak about the two planning models: "The essence is that if we try to enhance the quality of the public space and the safety of the people who use that public space, we hold on to a kind of ground-floor model, but we should also maintain the raised level of Hoog Catharijne." [123] Kernkamp agreed with Bakker:

> Bakker said "Hoog Catharijne is OK. It is a magnificent east-west axis. You should honor that. It was a very good concept at the time when Hoog Catharijne was built, a brilliant idea. It should be

maintained that way." When this was made clear, the City Council and the other parties considered it an eye-opener. We said "Let's please fix this town planning concept."[124]

Thus the city redefined the raised level as a "brilliant town planning concept." Situating the raised level of Hoog Catharijne in a tradition of daring architectural choices made it easier for the city to make this switch without loss of face. Hoog Catharijne had long been a kind of trauma for the city, but by using this argument Bakker provided a reason for the city to become proud rather than ashamed of it. Reconceptualizing Hoog Catharijne's design as not so bad after all implied that the city had not made such a huge mistake in the past.

Another crucial redefinition that took place at this point concerns the contrasts between the inner city and the Hoog Catharijne area. As was suggested above, the city had always wanted to make the boundaries between the inner city (old, open air, ground floor, small scale) and the Hoog Catharijne area (new, roofed, raised-level, large-scale) more fluid. At this stage, however, the city began to bring out the differences and dissimilarities between the two areas more clearly.[125] Since upgrading the ground-floor level was previously seen as the major strategy in diminishing the differences between the inner city and Hoog Catharijne, the aim of strengthening the contrasts decreased the importance of developing the area according to the ground-floor philosophy. Moreover, one feature of Bakker's plan that was crucial for the city was the proposal to return water to the Catharijne Canal.[126] This idea had an important symbolic value in the eyes of the city residents: it reminded them of the canals of the old inner city, an aspect of Utrecht that is greatly admired by locals and visitors alike.

Bakker's plan also included a radical renewal of the raised level, including the demolition of 25 percent of the Hoog Catharijne complex. This was possible partly because of WBN's redefinition of its shopping center. WBN started to make a distinction between Hoog Catharijne as a "run shopping center" and as a "fun shopping center."[127] Today Hoog Catharijne is still a "run shopping center." Many people "run" through it quickly. It is still the only relatively nice and safe way to get from the railway station to the old city center. These people do not necessarily buy anything. WBN's goal became transforming Hoog Catharijne into a center for "fun shoppers": people should come into the center and be persuaded to buy products by an attractive interior design and quality shops. Turning Hoog Catharijne into a "fun shopping center" would require radical remodeling.[129] This redesign plan started with the existing main raised pedestrian passage, the Radboud Passage. It was to be widened by 2.5 times its present width.[130] In this new mall concept, it was no longer necessary for everyone walking from the train station to the old center to pass through Hoog Catharijne. Thus, the possibility for a second main pedestrian route, via the inclined Station Square, was created.

Reconciling Technological Frames

In this phase of the process, redefinitions of goals and ambitions and new developments created new possibilities for the redesign of Hoog Catharijne. The two rigid technological frames that for so long had dominated the choices and interactions gradually receded into the background. And, not surprisingly, from this moment on the planning process seemed to make progress. Riek Bakker's success can be explained in part by the fact that she was an outsider to the nearly decade-long process in which two technological frames had dominated the interactions. It has been argued in the STS literature that outsiders are often needed to reject a dominant "mental model" or specific worldviews. Gorman and Carlson argued that, like technological frames, mental models may be a source of obduracy. They suggested that existing mental models can become so constraining that only a relative outsider can move beyond them and develop fresh ideas.[131]

   Bakker, without any prior stakes in either of the technological frames, succeeded in combining elements of both. Thus, actors do not always have to be either "closed in" or "closed out" of dominant technological frames. Bakker was "closed out" enough to be able to develop alternative designs and "closed in" enough to understand the backgrounds of the two technological frames. In this way, an "amalgamation of vested interests" occurred, one that is typical of stabilization processes in situations with two or more dominant technological frames. Combining the most important features of two frames involved strengthening both the ground floor and the raised level. In Bakker's plan the "ground floor" would be upgraded by filling the canal with water again and by reducing Hoog Catharijne's size by tearing down part of it and splitting the remaining structure in two, while the "raised level" would be enhanced by widening the Radboud Passage and building the inclined station square. Central Station would eventually become a separate entity again, and the Industries Fair would have better connections with Hoog Catharijne and the inner city. If the demands of the city, NS, WBN, and the Industries Fair were satisfied, they could consent to the plan without loss of face. Bakker's plan was approved by the City Council on December 18, 1997.

OBDURACY AND DOMINANT FRAMES

The planning concept on which Hoog Catharijne was originally based, a strict separation of traffic flows resulting in a design with raised pedestrian passages, had acted as a guideline for all planning proposals that were made for the area. Between 1987 and 1997 there were several attempts to reverse this concept. It was initially transformed into a model in which street-level pedestrian connections became the guideline: the "ground-floor model." Although the city and the railway company succeeded in linking this model to governmental policies and subsidies, statistics, and future prognoses, they did not succeed, in the end, in getting this model implemented. Some elements of Hoog Catharijne, the raised pedestrian passages in particular, were obdurate—hard to change or demolish. ABP vigorously tried to block every effort to transform these elements.

The concept of the technological frame and the distinction between closed-in and closed-out obduracy turned out to be useful in analyzing different types of obduracy for different groups of actors. The focus was on a major urban redesign process and on how, specifically, interactions at the local level between social groups, their constrained ways of thinking, and their strategies influenced the obduracy of Hoog Catharijne at various stages. Thus, the history of the UCP presents an example of obduracy caused by the dominance of certain "ways of thinking and acting" or technological frames. In the interactions among the partners, two technological frames were built up that became increasingly fixed. Because of their high inclusion, the city and ABP could not think of alternative designs outside the ground-floor frame and the raised-level frame. The rigidity of the opposing technological frames over the years led to a deadlock in the planning process. Only a limited number of alternative designs were proposed that either were in line with the raised-level frame or complied with the ground-floor frame. The clash between the two technological frames at this stage of the unbuilding process can also be stated in terms of power: Was ABP powerful enough to make the City Board accept its plans for the raised level of Hoog Catharijne? Residents, on the other hand, were closed out of the dominant technological frames and could only think of radical alternative designs, outside these frames. The closed-out obduracy experienced by citizens became visible in their proposals for radical changes of Hoog Catharijne, including demolition.

An outsider was needed to reconcile the two dominant frames. But not all outsiders were equally capable of rejecting the established ways of thinking. The architect Clemens Koemans and the citizens' group BOCP were outsiders who did not succeed in bringing about change. The fact that the citizens of Utrecht were more dispersed had negative consequences for their power position.[132] Riek Bakker did succeed however. Her advantage was that she was an expert who was in a position that enabled her to understand the backgrounds of the two technological frames and to exploit her strategic capacities in her interactions with the partners. Of course, Bakker could not have done this without being given power to influence the redesign process and the decisions. By "micropolitical means" such as lobbying, influencing actors' ways of thinking, and trying to bind actors to a common goal, she succeeded in opening up the "semiotic power structure," in terms of fixed meanings, that had become embedded in the opposing technological frames.[133] Several redefinitions of goals and ambitions and the presence of new actors in the planning process after 1996 thus increased the possibility of changing Hoog Catharijne.

Despite the initial successes of Bakker and her co-workers, residents of Utrecht remained critical of the project. Before the city elections in the spring of 1998, a new political party was founded. The main goal of this party, Leefbaar Utrecht (Livable Utrecht), was to prevent the execution of the Utrecht Center Project. Although Leefbaar Utrecht became Utrecht's largest political party after the elections, it did not join the City Board. In December 1999, the City Council of Utrecht had given its final consent to the new UCP plan developed under the supervision of Riek Bakker, but in March 2000 the four partners in charge of its implementation decided to end their cooperation, a decision that seems to have been triggered by budgetary problems.[134] Meanwhile, the city leadership and the railway company chose to form a new partnership aimed at implementing the UCP on the basis of the Definitive Town Planning Design.[135] Its major features regarding the redesign of the area's public space include the demolition of the Gilden quarter and the new Railway Station Square, the recovery of the Catharijne Canal, and the upgrading of the railway station to a "Mainport" public transportation facility. The two partners were seeking to enter into separate contractual agreements with the Industries Fair concerning the redevelopment of its premises, and with WBN concerning the redesign of the Hoog Catharijne shopping mall.[136] Although the partnership's breakup in March 2000 had already dealt a serious blow to the renewal effort, the chances of the plan's realization became further diminished as a result of the outcome of the local elections in the autumn of 2000. Leefbaar Utrecht, whose main goal was to prevent the execution of the Utrecht Center Project, gained some seats on the City Board after the elections. The City Board consulted the citizens of Utrecht about the new plans in a referendum in the spring of 2002. The majority of the people voted for a plan in which a more gradual connection between the Hoog Catharijne area and the inner city of Utrecht would be established. A new master plan was approved by the City Council in 2003. (As of early 2005, building activities have not started.)

Trying to Reconstruct the Highway That Cuts through Maastricht

The highway that cuts through Maastricht was built in the late 1950s. At that time, when cars were still comparatively sparse in the Netherlands, there seemed to be many good socio-economic reasons for building highways near downtown districts. Moreover, noise regulations did not yet thwart the construction of apartment buildings adjacent to highways. Between the early 1960s and the late 1990s, however, several interconnected processes radically changed the problems and practices of highway design: traffic increased dramatically; traffic safety and the quality of urban life became increasingly important issues; environmental concerns started to play a more important role in traffic projects; stricter environmental norms, regulations, and standards for the design of highways were developed; activists and lobby groups began to influence urban redesign projects; and local, regional, national, and international governments changed their policies on traffic circulation, in part because of other financial priorities.

Since the construction of the highway in Maastricht, there have been ongoing efforts by engineers, politicians, and citizens to change its design. It was clear from the beginning that the highway essentially split the city in two. Over the years, provisional adaptations were made, such as the reconstruction of intersections and the placing of sound baffles, but more radical new designs, such as overpasses or a diversion east of the city, failed to be adopted. The idea of a tunnel as the definitive solution to the congestion, the reduced traffic safety, and the poor quality of life of those who live near the highway has been considered and reconsidered for as long as 40 years. Although this idea has figured prominently in various municipal policies, proposals, and strategies, so far this solution is not even close to being implemented, nor is any other solution. Why is it that the highway that cuts through Maastricht has virtually remained unchanged for almost half a century? What were the strategies employed by the various actors engaged in the effort to improve the highway's design?

In this chapter, I analyze the efforts to redesign Maastricht's highway in the period from the early 1960s to 1998. Let me start by briefly outlining the context of the discussions in the Netherlands about the relationship between cities and highways in postwar society.

DESIGNING AND BUILDING HIGHWAY 75 (1956–1960)[1]

In the course of the twentieth-century, the increasing presence of cars in cities became an important aspect in urban planning. Postwar American cities, especially Los Angeles, have become exemplars of car-dominated places with busy freeways and traffic jams.[2] Many studies of the role of cars in cities have been dominated by technological determinist assumptions. But recently, urban historians have countered this point of view with more nuanced studies.[3] Eric Monkkonen and Clay McShane have argued that the twentieth-century "transport revolution" was not caused by the automobile but by a high-quality road system. Countering the standard view that the automobile developed "logically and almost inevitably from the invention of the internal combustion engine,"[4] McShane points out that the automobile technology emerged as a result of changes in American urban culture, and that the automobile, in turn, deeply influenced the configuration of cities. McShane relates the development of the automobile to changes in the public perception of the function of streets, arguing that initially streets were largely viewed as a meeting place, a recreational and social public space, but that after the arrival of the automobile this space became almost exclusively used as a transportation domain. Moreover, citizens came to view the automobile as an icon of liberation and an object of status and social prestige. This helped diminish the resistance of citizens to cars. The automobile was seen as a solution to various urban health and environmental problems associated with horses, including pollution, noise, and dust. It was seen as improving public safety because it could more easily avoid pedestrians. It was also seen as contributing to the realization of the ideal of suburban life and hence the liberation of the middle class. Significantly, McShane demonstrates how the emergence of the automobile gave rise to "utopian" plans for radical reconstructions of cities and how, on a more pragmatic level, it led to the widening and repaving of streets and to the building of bridges in cities. Both the "utopian" approach and the "incremental" approach assumed that it was possible to reconcile the existing design of cities with the challenges and intrusions of automobility.

After World War II, Dutch planners, road engineers, politicians, and policy makers began to think about the future of cities in relation to the expected increase in the number of cars."[5] In 1948 the Vereniging het Nederlandsche Wegen-Congres (Association of Netherlands Road Congresses) organized a conference about the question whether highways should be built *in* or *around* cities. Highways with overpass junctions had already been introduced in several major cities in the United States, where initially they were mostly built through poor urban areas on the pretext of "slum clearance." In the 1950s, when the urban freeway-building program was at its peak and middle-class neighborhoods began to be affected, there were protests, and local activist groups were established to stop these projects.[6]

One of the speakers at the 1948 conference, L. H. J. Angenot, director of Rotterdam's planning department, observed that such massive highway projects were still absent in Europe. Traffic congestion was not seen as a serious problem yet in European countries. Moreover, in the years immediately following the war, funding went to projects that had more urgency than a state-of-the-art highway system. It seemed nevertheless obvious that automobile traffic would significantly increase in the years to come, and Angenot suggested that it was necessary to begin thinking about the implications of this growth for cities and highway construction, and whether or not in this respect the American example should be followed.[7]

The conference's main concern was how to reconcile the growing number of cars with the existing and future layouts of cities.[8] G. C. Lange, another planner, argued that traffic congestion in cities might result from the difficulty to adapt roads. Over the years, houses in old neighborhoods had been replaced, but the roads along which they were built often remained unchanged for hundreds of years. This meant that minor adjustments in width, in trajectory, or in the number of crossings might still require radical and hence costly alterations. Furthermore, Lange argued, the design options for roads and streets in cities generally were quite limited; road curves in particular were much easier to design for the spacious countryside.[9]

In the 1950s, Maastricht, like many other Dutch cities, had to address questions such as "Where should the new highway be situated in relation to the city center?" and "How should it be designed?" The City Board, in collaboration with Rijkswaterstaat (the national governmental body for road construction and road management),[10] had considered plans for a north-south highway near Maastricht before World War II.[11] Although Rijkswaterstaat initially projected the highway to be east of the eastern districts Heer and Amby (farther from the center of Maastricht), eventually it was planned to be closer to Maastricht's downtown area.[12] The Chamber of Commerce of South Limburg effectively lobbied for a road near the city's main industrial zones, which happened to be adjacent to the downtown area.[13] The companies, of course, would benefit from enhancement of the nearby infrastructure. Moreover, it was argued that a new highway farther from the city's center would barely contribute to the local traffic circulation.[14]

The City Council of Maastricht was well aware of the new highway's significance to the city: "There is no doubt that Highway 75 is of utmost importance for Maastricht. This era's traffic development depends on roads and highways rather than—as in the old days—on rivers and railroads, and now that Highway 75 passes Maastricht it is important to effectively integrate it into the city's traffic system."[15] The City Council also emphasized the necessity of good roads for the future development of Maastricht as an industrial center and underscored the significance of having a highway that connects the city to other parts of the country. It believed that for Maastricht, "as a city at the intersection of cultures, good and fast connections are necessary. . . . When this highway is completed, Maastricht will eventually have a good and fast connection with the central and western part of the country."[16]

The director of the city's Public Works Department, J. J. J. van de Venne, argued that Maastricht fulfilled a major socio-cultural function for the region.[17] He felt that the new traffic plan should facilitate the fluid passage of international traffic through the city and that it should enable regional traffic to come as close as possible to the city's industrial areas and its public services and institutions. At that time, it was expected that Maastricht would become centrally located between two main highways for which plans were underway: the highway between Antwerp and Aachen via Liège, south of Maastricht, and the highway between Antwerp and Aachen via Elsloo/Heerlen, to the north of Maastricht.[18] Van de Venne also referred to the competition Maastricht might have to face from the nearby Dutch cities Heerlen and Geleen: "If Maastricht wants to maintain its socio-economic influence and successfully compete with new developments like those in Heerlen and Geleen, the city will absolutely have to pursue expansion and reconstruction . . . in the near future."[19] Van de Venne perceived a direct link between the city's socio-economic ambitions and its traffic problems, and argued that the inner city, with its commercial district, should be accessible for cars by means of a circular system of roads.[20] The projected ring road was seen as the backbone of Maastricht's traffic-circulation system, and the urban stretch of Highway 75 constituted a part of it[21]: "One of Maastricht's most important entranceways is the E-9 motorway or Highway 75. The urban stretch of this highway was under construction from 1958 to 1959 and forms part of the traffic circulation around the inner city."[22]

Despite the enthusiasm, City Council discussions in 1958 reveal the doubts of some of the council members about the chosen trajectory. In one meeting, a council member named Wishaupt wondered if it would not be better to build the highway on the outskirts of the city with branches leading into the downtown area.[23] He also objected that the planned Highway 75 was part of plans for the extension of Maastricht that had already been ratified a few years before.[24] Therefore, he claimed it was not possible to propose important adaptations. With reference to previous experiences in the central part of the Netherlands, it was argued that the building of highways around cities had obvious disadvantages. The chairman of the council mentioned the example of Utrecht, where the highway allowed all through traffic to bypass the city.[25] It was thus eventually decided that the highway would cut right through the eastern part of Maastricht, just east of the Maas River and the railroad tracks.

The highway engineers of Rijkswaterstaat may have had doubts about the suitability of this trajectory, because formerly it had been the site where the overflow of water from the river ended up. As the Rijkswaterstaat highway builder Jacques Jamin suggested in an interview, "an old saying in road engineering is that a road always needs to have dry feet."[26] It is therefore remarkable that the new road was planned in what used to be a very wet area. Because of the site's history as an overflow reservoir, it was a very low-lying area, with a comparatively high groundwater level. This concern, however, would not reverse the plans. Clearly, the economic and socio-cultural arguments of the city leadership and the Chamber of Commerce prevailed and fixed the trajectory of Highway 75 at its present site: right through the city.

The Design of Highway 75

The planning and building of Highway 75 took place between 1956 and 1959. In 1948, L. H. J. Angenot, speaking at the conference of the Vereniging Het Nederlandsche Wegen-Congres, had described three ways to connect a highway to a city.[27] The first model involved a highway in or near the city and a ring road or beltway outside the city. In the second model, called a "linear system," main highways were situated at a relatively large distance from cities; highway and city were to be connected by a network of smaller highways or secondary roads. In the third model, the "passage system," highways were led through the city. Angenot pointed out that the first and the second system were most common in the Netherlands at that time. In contrast to American cities, where highways often penetrated densely populated urban centers, European countries favored the ring road system. In the late 1940s, then, the passage system was very uncommon in the Netherlands. Angenot argued that it often channeled regional and national through traffic and also affected the local traffic circulation. Therefore he suspected that the passage system could easily cause congestion. In American cities, a ring road was often added to a highway to diminish the inconvenience that local traffic caused for through traffic.[28]

The way Highway 75 was linked to the city of Maastricht most resembled Angenot's third model: the passage system. Initially the plan was to construct overpasses at two of the main intersections with local roads: Geusselt and the Scharnerweg.[29] The city supported this solution. In June 1958, some council members voiced their concerns about the level road junctions, especially with regard to the safety of school children that had to cross the highway. Moreover, they argued that the new highway would add a third north-south barrier to a city that was already divided by the Maas River and the railroad tracks. This is why they favored building a highway below ground level. Rather than opting for overpasses to allow for safe intersections, they favored lowering the new highway to achieve that same effect.[30]

The Ministry of Transportation subsequently postponed the construction of overpass junctions.[31] This meant that for the time being the intersections would remain at ground level. It was projected, however, that in the near future Highway 75 would pass below the Scharnerweg, a major intersection close to the Railway Station, while it would be elevated near Geusselt, another major intersection a kilometer to the north of Scharnerweg. At a later stage, the highway would ultimately pass below both intersections.[32] This was the basic scenario in 1964, when J. J. J. van de Venne claimed that "in the near future" the highway passage would be without level road junctions altogether, and only two access roads would link the highway to the city.[33] In 1969, the author of the city's yearbook reported that the plans for realizing an urban highway without level road junctions have reached "an advanced stage."[34] The author mentioned four intersections along the highway's urban stretch that, according to these plans, would be turned into overpass junctions.

Another option that was put forward in the late 1950s was the idea of building a tunnel for Highway 75. One of the proponents, council member Schreuder, argued that Maastricht would "only benefit from the immediate construction of a tunnel."[35] The author of the 1958 Maastricht yearbook also referred to this issue: "It is the intention of the government to build a tunnel at this place in the future."[36] Between the roads to and from the Geusselt traffic circle, space was set aside in order to realize the entrance of the tunnel at some point in the future.[37] However, in retrospect, Rijkswaterstaat engineer Jamin disputed the assumption that a tunnel would be built at that time. He suggested that the confusion about the tunnel was due to the fact that many people failed to make a clear distinction between a tunnel and a lowered road, indicating that some articulated ideas about building a lowered highway with overpass junctions, not a tunnel.[38]

Whereas a lot of space was reserved near the Geusselt traffic circle on the northern edge, at the southern edge of the urban section of Highway 75 much less space was set aside: the highway was in fact planned immediately adjacent to Maastricht's first high-rise building, the "Municipal Apartment Building," which was built between 1948 and 1950.[39] The building, designed by city architect F. C. J. Dingemans (1905–1961) and consisting of 90 apartments, was built in response to the great shortage of homes in the postwar era. From the beginning its design included elements that were seen as very modern at that time, such as elevators, central heating, and refuse chutes.[40] In general, it seems, the city's leadership viewed this building, and the other apartment buildings that were planned to line the urban stretch of the highway, as showcases of the city's modernity. Consequently, a lowered highway with slopes was not an option. It would take up too much space and obscure the buildings from view.[41] Rijkswaterstaat, in contrast, preferred a spacious design and a less conspicuous presence of buildings along the highway. It argued that having a more spacious layout of the highway would "prove to be the right opinion in the

Figure 3.1
The Municipal Apartment Building in 2000. Photograph by A. Hommels. © A. Hommels.

Figure 3.2
The Municipal Apartment Building in 2000. Photograph by A. Hommels. © A. Hommels.

future."[42] Notwithstanding this view, the apartment buildings were eventually placed relatively close to the highway.

Those in charge of the decisions involving the highway's construction in the 1950s deliberately tried to keep various redesign options open, but, ironically perhaps, in later periods it turned out to be extremely difficult to radically deviate from the original design. The ultimate plan was based on a lot of considerations: opinions about traffic engineering, views of the role of automobiles in cities, socio-economic considerations, planning motives, experiences in other Dutch cities, and earlier extension plans—all became integrated and embedded in the design of Highway 75. This underlines the importance of understanding the highway as part of a larger socio-technical structure, consisting of the highway itself, the city's traffic-circulation system, the apartment buildings surrounding it, and the many ideas and implications involved in its design. As will become clear in the remainder of this chapter, many features that were built into the highway's design maintained their obduracy for quite some time. At what points in the process did actors still have concrete opportunities for changing the highway's design? In the remainder of this chapter, I will analyze how, under the influence of new ideas and developments, the socio-technical structures of the urban section of Highway 75 became contested.

The effort to unbuild the urban section of Highway 75 can be divided into four periods, each marked by a significant shift in the types of unbuilding activities that were undertaken. The first period began in the early 1970s when Rijkswaterstaat and the city sought to better integrate the highway into the overall urban traffic scheme by reconstructing the highway's main intersections, reconstructions which eventually took place between 1974 and 1978. The second phase covers the years between 1978 and 1982, when a Trajectory Study was done in which two redesign options played a major role: a tunnel and a diversion east of the city. This study caused decisions about the implementation of a possible solution to be postponed, thus triggering a new phase, lasting from 1982 until 1993, when various efforts focused on temporary measures. Finally, the years between 1994 and 1998 once again reflect a more radical effort to reconstruct the highway, mainly because of the Trajectory/EIS Study.

RECONSTRUCTING THE INTERSECTIONS (1974–1978)

By the 1960s, city planning and urban road design had become increasingly linked in the Netherlands. In a 1965 article in *Wetenschap en Samenleving* (*Science and Society*), H. A. M. C. Dibbits underscored the significance of their integration. Specifically, he pointed to the difficulty of dealing with historically grown situations in towns, where buildings of cultural and historical importance could not be simply demolished to allow space for new roads. Dibbits also warned for the cumbersome nature of traffic arteries and suggested that in many situations it was impossible to fit them in afterwards.[43] Although there seemed to be some agreement that old inner cities had to be preserved, many Dutch cities initiated large-scale reconstructions in the 1960s. This involved, among other things, new ring roads around downtown districts, beltways around cities, and the paving over of canals to make space for cars.[44] It was also deemed important, mainly for reasons of safety, to separate the various types of traffic. Thus city planners were faced with the challenge of putting in place two different systems side by side: one for motorized traffic (cars and buses) and one for slow traffic (pedestrians and bicyclists).[45]

The completion of the urban stretch of Highway 75 did not mute the debates about its design. Almost from the start, the residents of Maastricht began to voice their concern about structural flaws. By the early 1960s it was obvious that the urban section of the highway was quite unsafe. One of the main intersections, the Scharnerweg, seemed to baffle many drivers. The introduction of traffic lights at this intersection significantly improved the situation,[46] but in time other reservations were put forward that had to do with the delays that resulted from the increase of local traffic and with the general unsafety of the highway. Moreover, complaints about the noise produced by the crowded highway grew louder, while concerns about its barrier function, dividing the city in two, resurfaced. Soon the Province of Limburg, the residents of Maastricht, the city leadership, and Rijkswaterstaat agreed that the highway's design had to be changed more radically.

The Highway's Integration in Traffic Systems and Planning

During the 1960s, the road had become more and more incorporated into the local and international traffic-circulation network. In 1970, a cloverleaf at the Europaplein was constructed, situated at the southern tip of the highway's urban section.[47] From there on, the highway extended southward about ten more kilometers before it fed into the Belgian highway system.

The construction of this final stretch linked Highway 75 with the international highway network. In addition, the highway's integration in the local traffic network was further consolidated when just to the west of the highway, near the Scharnerweg intersection, a tunnel was constructed, which facilitated a better flow from the eastern part of the city to the downtown district.

In the course of the 1970s, traffic on Highway 75 further increased. The authors of a 1974 Rijkswaterstaat report pointed out that the highway's urban section and the adjacent buildings were built in an era when no one anticipated serious traffic growth and traffic nuisance hardly was a factor.[48] But meanwhile the capacity of the highway had proven insufficient, the road surface was in bad shape and the traffic lights regulation system was no longer up to date. As a result, the authors argued, a dangerous situation had arisen that was especially unacceptable to the people who lived nearby.[49]

Rijkswaterstaat and the City Board decided to reconstruct the road in 1976. In August 1977, the Geusselt traffic circle was transformed into an intersection secured by traffic lights. According to Rijkswaterstaat, this adaptation was necessary to improve the flow of traffic and to avoid future congestion. One of the other goals of this reconstruction was to improve the safety of cyclists and pedestrians. According to representatives of the Province and road users, the changes were not implemented quickly enough. In 1977, the Province began to put more pressure on Rijkswaterstaat to improve the safety of the highway. Meanwhile, parents whose children attended the primary school near one of the intersections challenged the City Board for poor regulation of the traffic lights.[50] Some of these parents protested by blocking the highway.

One of the main difficulties in the highway reconstruction process involved the Scharnerweg intersection. Rijkswaterstaat analyzed the various options and found that the intersection's main problems comprised its low capacity, its danger for pedestrians and other slow traffic, the high noise levels, and the continuous delays for public transportation when crossing the highway.[51] The options that were mentioned earlier cropped up as solutions once again: an overpass junction or a level road junction in combination with a diversion east of Maastricht. In the choice between these redesign proposals, the integration of the highway in the larger infrastructure and traffic scheme of the city that had been established in the late 1950s played a crucial role. Furthermore, user practice became an important consideration in the choice for redesign options. It turned out that the volume of local traffic on the urban section of the highway was larger than the volume of through traffic. As a result, the diversion was rejected because experts thought it would hardly reduce the density of traffic on the urban stretch of 75. Moreover, it was argued that planning this diversion would take up a lot of time and money. Similarly, the option of an overpass junction was believed to be extremely expensive. Besides, experts anticipated that the new bridge across the Maas River that was already planned, the Noorderbrug, would alleviate part of the congestion on 75, so that an overpass junction at the Scharnerweg would, in effect, no longer be necessary.

The city leadership emphasized the importance of the urban highway stretch for local traffic circulation. In 1964, Van de Venne already claimed to be pleased the road did not function as a full-blown highway, because that diminished its barrier function as well. Moreover, he argued, the highway's current design allowed it to be in integral element of the ring road around Maastricht, in part because of the highway's multiple connections with the urban traffic network.[52] The city leadership underscored the highway's significance for local traffic, specifically as access to the downtown district. If its integration into the local traffic network would be omitted, the downtown area would be much harder to reach by car. Therefore, the city argued, overpass junctions *without* easy connections to the local infrastructure of roads would be unacceptable. In this way, the view of the city leadership contributed to the obduracy of the existing design, including its connections to the inner city.

Another reason why it was difficult to reconstruct the Scharnerweg intersection had to do with its function in the larger planning structure. In the 1950s, the city leadership deliberately opted for the highway's inclusion in a representative planning structure, including modern apartment buildings. After the construction of these buildings along the highway it became increasingly difficult to implement changes. To the City Board, a solution with overpass junctions was unacceptable, because it would involve rigorous adaptations of the intersections and the demolition of buildings to find space for access roads and exits.[53] The preservation of the apartment buildings and the wish not to spend much money on a solution played an important role for the city leadership. This illustrates that obduracy has a distinct financial side to it: once urban structures have become embedded it can simply be very expensive to "unbuild" them.[54]

As a result of these considerations, only ground-level solutions were acceptable: the highway was slightly diverted westward, so it would be a little less close to the existing apartment buildings. This limited the noise somewhat, in particular for the residents of the Municipal Apartment Building, which was closest to the highway. Furthermore, Rijkswaterstaat installed a new traffic lights system.[55] Although the City Board was not altogether satisfied with the reconstruction plans, they thought that doing nothing was worse; moreover, the new measures were considered to be only temporary.[56] In June 1977, the reconstruction of the Scharnerweg intersection neared completion. In 1978, the traffic lights were phased for 50 kilometers per hour to ensure a fluid passage of cars through Maastricht.

Embeddedness as an Explanation for the Highway's Obduracy

So far, I have analyzed how the urban section of Highway 75 became integrated in larger traffic schemes, user practices, and planning structures, each factor further complicating the possibility of radical changes in the highway's design. As my analysis suggests, material factors alone could not account for the highway's

Figure 3.3
Reconstruction of Scharnerweg intersection (1976). Source: Gemeentearchief Maastricht.
© Fotopersbureau Widdershoven.

obduracy. I demonstrated, for instance, how the commitment to existing planning structures, financial arguments and user practices played a crucial role. Neither was it useful to explain the complexities of redesign with reference to the existence of dominant frames. At this stage, the actors did not rely on dominant technological frames that guided their interactions and ways of thinking. The redesign options, in other words, were not limited by a particularly high or low inclusion of actors in a specific technological frame. This also explains that the sharp conflicts that frequently result from strong commitments to technological frames, as was the case in the Utrecht City Project, were hardly a factor. The highway's resistance to change at this stage of the process can best be explained, I believe, by the specific forms of embeddedness I described: the highway's embeddedness in the local, national, and international traffic system and its integration in Maastricht's planning structure. In what I identified as the second period in my account, the phase of the Trajectory Study (1978–1982) I will discuss the role of the tunnel option prior to this period a well as afterward, when this idea became more and more integrated in various relevant urban spatial policies.

TRAJECTORY STUDY (1978–1982)

By the 1970s, policy makers' opinions about the location of highways in relation to cities had become more pronounced: highways should preferably not go *through* cities (where they cause all kinds of environmental, safety and livability problems), but *around* them. This shift took place in the context of a more general change of opinion among planners, policy makers and citizens about large-scale urban interventions in the 1970s.[57] Whereas in the 1960s increased automobile traffic often served as a legitimization of drastic solutions, including major interventions in infrastructure (filling in canals) and urban districts (demolishing old towns) that had been around for centuries, this was no longer accepted in the 1970s. Local activist groups increasingly voiced loud protests against huge urban redesign projects.

In the late 1970s, three policy reports mentioned the possibility of a diversion of the urban section of Highway 75 (now named A2) around Maastricht.

The Province of Limburg proposed the diversion as an option, but stated that putting in overpass junctions at the present trajectory still had priority. The national government's 1976 Structuurschema Verkeer en Vervoer (Structural Plan for Traffic and Transportation), a general planning report that summarized the most important policies of the Ministry of Transportation, also referred to a diversion of the highway in Maastricht, to be completed in the 1990s. The report argued that highways of the main national network should bypass urban areas, rather than go through them, because the absence of large flows of traffic in cities would improve urban living environments and reduce risks such as those associated with the transportation of hazardous materials. These policy reports caused the city's leadership to decide to set aside a strip of land that in the future might be used for the diverted highway east of the city. As a consequence, this area could no longer be used for other extension plans of the city. The municipal Structural Plan mentioned the possibility of a diversion of the highway east of Maastricht, but it also anticipated urban expansion at the eastern edge of the city. Although there was not very much (international) through traffic on Maastricht's highway yet, the City Board considered the prospect of heavy trucks passing through the city 24 hours a day as unacceptable.[58] Because traffic was expected to go up and international standards for the design of highways were ever tighter,[59] the City Board felt pressured to investigate the possibilities for adaptation of the existing trajectory or its diversion.[60]

After World War II, Maastricht, like many other Dutch cities, suffered from lack of space and scarcity of decent housing. In 1962, the director of Public Works, Van de Venne, had already proposed the annexation of neighboring villages, to establish a "Greater Maastricht."[61] The planning effort aimed at the city's expansion was affected, of course, by the uncertainty about the highway's future. Although the mere mention in policy documents of a diversion on the eastern edge of the city implied a claim on this area, other actors advanced competing proposals for this area. These claims involved urban expansion, nature reserves and water procurement areas, as well as other possible road connections east of Maastricht.[62]

Because of the projected traffic growth, changed political and urban spatial policies, and international regulatory adaptations, discussions about the highway's design resurfaced. It was decided to study the various trajectory and design
alternatives so that the City Board could make an informed decision on how to
use the available land. The project team "E9 and Maastricht," that was going to
be in charge of the Trajectory Study, was formally established on June 16, 1978
by the mayor and City Board of Maastricht and the chief engineer and director
of Rijkswaterstaat.

This team studied two design variants and a number of subvariants. One
variant started from the existing trajectory of the highway and explored a closed-
tunnel option, a lowered-highway option, and a lowered-highway-with-
acoustic-fencing option. A distinction was made between designs that either did
or did not integrate the Scharnerweg intersection. The second variant explored
possible diversion trajectories along the eastern edge of Maastricht.

As was pointed out above, the idea of a tunnel to resolve the problems
associated with the urban stretch of the highway was mentioned already in the
late 1950s—in the annals of the city as well as in the minutes of a City Council
meeting. In the 1970s, when Rijkswaterstaat and the city were thinking about
reconstructing the highway's intersections, the idea of a tunnel turned up again.
Residents of the nearby apartment buildings established an action committee,
called E-9 Ondergronds, in which members of environmental activist groups
also participated.[63] In a letter to the City Council, the committee argued that the
plans for reconstructing the Scharnerweg intersection were unacceptable. In
their view, the planned reconstruction would not contribute to a reduction of
the noise produced by the highway. They claimed that a tunnel could be built
without demolishing any houses, while at the same time keeping all the existing
connections of the highway to the local traffic network in place.[64] Moreover,
they argued that the proposed modifications would make future adaptations
increasingly difficult.[65]

Although a tunnel would be more expensive than the diversion, the city's
leadership also preferred a closed tunnel. By the early 1980s, a connection to the
Scharnerweg was no longer deemed necessary by the city[66]: "The only right

solution in our view is a tunnel at the location of the present trajectory, especially now that its technical feasibility has been established."[67] Since then, the city's leadership has been fairly consistent in its preference for a closed tunnel.[68] According to P. Jansen, a city traffic engineer, the city opposed all lowered road variants because none of them could sufficiently diminish the various forms of environmental damage produced by the urban highway stretch.[69] Only the closed-tunnel option would meet the environmental standards (those regarding noise pollution in particular), while it also significantly reduced the highway's barrier function, something the city strongly supported.[70]

The City Board rejected a diversion east of the city. A diverted highway along the city's eastern edge would form a barrier as well and would completely do away with the city's smooth transition into the surrounding landscape.[71] The diversion would negatively affect agriculture, drinking water supplies, recreation and the living environment of the residents of the eastern districts. Moreover, a diversion would conflict with the then prevalent "city border philosophy" of the city's leadership that aimed at a gradual transition from city to countryside.[72] This philosophy had become increasingly important and was closely linked up with the city's identity. Former City Manager Ad Lutters commented that in the eyes of the city leadership "the eastern border has been a sacrosanct fringe that should not be touched," a view that almost became "a kind of dogma" over the years. Situated in a river valley and surrounded by rolling green hills that contain the characteristic yellow marlstone, Maastricht has always cherished its unique geography.[73]

Apart from the traffic and cost arguments put forward in the discussions about the reconstruction of the intersections in the mid 1970s, environmental arguments (landscape, water supplies, vistas) and the importance of a gradual transition from city to countryside were now advanced to discredit the diversion option. In the opinion of both the city leadership and the residents the city's eastern border was rigid—it should not be tampered with.

The idea of a diversion had nevertheless played an important role in Rijkswaterstaat's way of thinking for a long time; it had already been considered—and abandoned—twice: once before World War II and once in the mid

1970s. Although the amount of traffic did not necessitate measures at that time, Rijkswaterstaat opted for a diversion around the city for reasons related to traffic growth, traffic safety and the living environment. In Rijkswaterstaat's view, safety was paramount and therefore it did not favor a closed tunnel: the transportation of hazardous goods through tunnels was widely considered to be too risky. Furthermore, its wish to transform the urban section of the highway into a "real" highway could be easier realized by opting for a diverted trajectory. In general, highway engineers find it easier to build highways in relatively open spaces than in built-up urban environments, where existing planning structures seriously limit the available design options.[74] A Rijkswaterstaat memo of April 1978 suggests that at that time the decision process was primarily geared toward the diversion variant rather than the existing trajectory.[75] Even though this preference was denied in an earlier letter to City Board member Dols dated January 27, 1978,[76] in retrospect Jamin (a highway engineer involved in the design process) emphasized the importance of the diversion alternative for Rijkswaterstaat: "At that time, we were not at all working on a road trajectory that went through the city. We were only studying alternatives outside the city. At the very last moment, we received a directive from Den Haag: 'You should also make a plan for a highway that cuts through the city.' There were, however, only broad outlines for such a plan, while the diversion option, by contrast, was thoroughly investigated, including zoning schemes."[77]

In 1981, the residents of Maastricht were publicly consulted about the proposed diversion. A majority appeared to be against the diversion, whereas most considered a tunnel below the existing trajectory the best solution.[78] According to the residents of the eastern city districts, the diversion of the highway would locally devastate the environment. Yet the residents of the apartment buildings along the existing highway viewed their situation as "unbearable." The smell of exhaust fumes and traffic noise made their lives utterly miserable. Remarkably, perhaps, the majority of them also viewed a tunnel as the best solution. The old highway would then become superfluous, they argued, and should be transformed into a green area or park.[79]

The Raad van de Waterstaat (Council for Water Works) had to advise the Minister of Transportation about the Trajectory Study. They reached this conclusion: "Reconstruction of the present passage through Maastricht is, when taking into account nearly all aspects, more attractive than the construction of a diversion around the city."[80] On January 13, 1982, Minister of Transportation H. J. Zeevalking (a member of the D'66 party), decided in favor of the trajectory through the city. He did not say when the solution should be implemented, nor did he stipulate which of the three tunnel options should be built—a closed tunnel, an open tunnel, or a semi-closed tunnel with acoustic fencing.[81] Unambiguously, however, the minister stated that a diversion was not a good option.[82]

The hesitation to build a tunnel immediately can be explained by a number of interrelated cultural, historical, economic and political factors. In 1982 there was no congestion on the urban section of the highway. The oil crises of the 1970s had culminated in economic recession by the early 1980s. This, in turn, had diminished the growth of traffic, thereby decreasing the need for immediate action.[83] Moreover, in the 1970s, Dutch people had become more aware of environmental problems; automobile usage was increasingly criticized for its contribution to air pollution and the depletion of fossil fuels. Governmental transportation policies became more focused on reducing the number of cars. In addition, due to the then current noise regulation standards and tight budgets, full realization of the tunnel within a short time span could not have been possible.[84] Furthermore, the problems of Maastricht's highway infrastructure were considered minor in comparison to those in the more densely populated western part of the Netherlands. Finally, at that time Maastricht and the Province of Limburg simply failed to have a strong highway lobby in Den Haag, the seat of the national government.[85]

After the 1982 decision of the minister, the diversion option stopped playing a role: future alternatives would be defined in line with the idea of a tunnel and were limited to a reconstruction of the existing trajectory. The decision has since become more fully integrated in various municipal policies; as a result, a tunnel became a must in the eyes of the city. Clearly, the minister's decision contributed to the tunnel's further embeddedness in the city's plans and policies.

After 1982, the idea of a tunnel became more and more part of the expectations and ideals of residents and local politicians alike. Rijkswaterstaat and the city began to work on several tunnel variants in the Working Group Tunnel Design that was established in 1989.[86] For a while, the city strongly opted for an expensive drilled tunnel, since this was seen as the only possibility to avoid large-scale demolition. In the early 1990s, local real estate agents sold houses in nearby neighborhoods claiming that the city highway would soon disappear in a tunnel with the "old" highway transformed into a city park. Residents of Maastricht also unfolded initiatives aimed at developing tunnel plans. In the mid 1990s, the local planner René Daniëls, director of Buro 5, proposed a tunnel design, to be developed in collaboration with project developer HBG Vastgoed and an engineering business. Their provocative plan played an important psychological role in the overall planning process.[87] According to Daniëls, the plan was largely in accordance with the city's viewpoints in the 1990s; the soon-to-be former highway, for instance, was designed as a boulevard with trees. Moreover, demolition of apartment buildings would not be necessary in Daniëls's solution.[88] Although Armand Cremers, the City Board member in charge of traffic issues at that time, liked the scale model of the plan very much, the plan was never seriously considered as a definite solution by the city. Looking back, Cremers argued that the plan turned up "too early" in the process of thinking about solutions and that it was "too concrete." He suggested the solution did not fit in the existing planning structure and failed to solve the financial problems.[89]

Embeddedness of the Tunnel Option in Spatial Policies, Ideals, and Expectations

It is perhaps surprising that the tunnel option, despite its popularity with the residents and leadership of Maastricht, is still not implemented. The idea of a tunnel played a central role in the expectations and ideals of those immediately involved. In time, the tunnel idea acquired robustness: it became difficult to neglect in the unbuilding process. Obviously, its obduracy had no material component, for the tunnel did not exist as a reality yet. The tunnel grew obdurate, in

Figure 3.4
"Hideaway" tunnel plan for highway through Maastricht, by Buro 5. © HBG Vastgoed and
Buro 5 Maastricht.

the first place, because it became linked with expectations and ideals of residents: already in 1958, the tunnel idea was mentioned. That same year, council members proposed a tunnel as the preferred future design of the urban highway stretch. In the 1970s, residents who were represented in the E-9 Ondergronds action committee made a strong case in favor of a closed tunnel. The public reactions to the 1980 Trajectory Study showed that residents preferred a closed tunnel with a park on top of it. Most recently, the tunnel turned up in a new citizens' initiative: Buro 5 designed a tunnel and visualized it in a scale model. Second, the idea of a tunnel acquired more robustness when it became integrated in municipal policies and activities: the city established a working group, particularly aimed at designing tunnel variants. After the decision of the minister in 1982, the idea of a tunnel in Maastricht became further integrated in local plans and policies. [90]

I have now made a paradoxical move by using the concept of embeddedness to explain the obduracy of a technology that is not even implemented yet. I argue that particular solutions, as in this case the tunnel option, can become embedded in urban, regional and national policies, as well as in the ideals and activities of citizens and politicians, *before* they even become a physical reality. Thus embedded, the tunnel, while non-existent, had important effects on the unbuilding process.

TEMPORARY MEASURES (1982–1993)

After the minister's 1982 decision, the diversion alternative was no longer discussed. Although the tunnel had become the only serious option, it nevertheless receded to the background because its implementation was postponed. The tunnel was increasingly seen as a project that would be realized in some distant future and this opened up debates about short-term measures, for, after all, most of the highway's problems did not simply vanish. Because actual tunnel construction was not planned to begin before the 1990s, temporary measures were considered necessary to lessen the nuisance caused by the highway. The decade of the 1980s was one in which highway design in urban areas in general, and the urban stretch of the highway in Maastricht in particular, became increasingly tied to legal standards and regulations. This made it more difficult to make major adaptations in the highway's design. During that period, environmental laws on noise and toxic emissions gradually toughened.[91] In Maastricht, a range of measures aimed at reducing the traffic noise for the nearby apartment buildings (the milieubouwplan) was executed in 1982.[92] In 1983, residents protested the sound baffles that were built, claiming that they made the highway less safe: "As long as we have no tunnel with overpass junctions, the risk of accidents remains. All temporary measures you take are palliatives. Much money has been spent on reconstructions already, but the highway hardly became safer."[93] Residents protested vehemently against the building of a wall to lower the traffic noise,[94] even comparing it to the Berlin Wall.[95] The protests, however, led only to a few adaptations in its design. Three years after the acoustic fencing was put in place, the residents of the Oranjeplein apartment buildings were still all but satisfied. They argued that the wall isolated the neighborhood behind it, impoverishing the area. Moreover, traffic safety had gone down and traffic noise had aggravated rather than diminished, according to the residents.[96]

It was clear, however, that the reconstruction of the highway in Maastricht was not a large priority on the national agenda. In the Second Structural Plan for Traffic and Transportation (Tweede Structuurschema Verkeer en Vervoer 1988), the A2 was labeled a "main transport axis" instead of a "hinterland connection."[97] As such, it was given lower priority, which meant that no budget was set aside for its reconstruction.[98] The City Board, the Province of Limburg, and the Chamber of Commerce were worried about this and decided to join forces in their search for temporary solutions.

On July 12, 1991, a Working Group for Temporary Measures was established, consisting of members of the city, Rijkswaterstaat, the Province of Limburg and the Chamber of Commerce. Since it was expected that a tunnel could not be realized before 2000, temporary measures were seriously studied. These involved the adaptation of the traffic regulation scheme, adaptation of the Scharnerweg intersection, and broadening the highway. Since an increase of traffic was expected, the measures were mainly geared towards avoiding congestion.

Figure 3.5
Protest against sound baffles, President Rooseveltlaan, 1983. Source: Gemeentearchief Maastricht. © A. Werker.

In this phase a new alternative solution emerged, the "solution of Liège." This meant that the local roads would pass above the highway at ground-floor level. The Chamber of Commerce of Limburg, representing the interests of the regional businesses and companies, advocated this solution since it would be a quick solution that could be built within short term. For the Chamber of Commerce, the improvement of the accessibility of Maastricht and the region was the most important goal to be achieved. Quick, provisional solutions, it believed, should be favored over long-term solutions.[99] However, some of the solutions proposed, e.g. the solution of Liège and a widening of the road to six lanes, were not possible because of the 1979 Wet Geluidhinder (Noise Abatement Act). This act put limits to the amount of noise produced in specific "noise zones." Within such a zone, for instance, houses can only be built if the norm of 50 decibels is not exceeded. In urban reconstruction projects the norms are somewhat less strict.[100] Paradoxically, the Noise Abatement Act served both as an instigator of change and as a conserving force. On the one hand, it enabled taking measures in zones in which the noise levels were too high; on the other hand, it prevented implementing new building or reconstruction options in other cases, thus contributing to the obduracy of the existing structures in those areas.

In 1992, the European Commission issued a directive on environmental impact assessment that made an Environmental Impact Statement (EIS) mandatory for highway construction. Because the 1982 decision on the urban highway section in Maastricht had not yet been implemented, this new procedure also applied to that decision.[101] Moreover, the 1993 Trajectory Law[102] stipulated that the 1982 decision had expired because it had not been implemented within a 10-year time span. Thus, new laws and regulations had caught up with the earlier decision to change the highway's design. Whereas the implementation of temporary measures did not require an EIS procedure, an adaptation of the local zoning scheme (defined by law) would be necessary. In a zoning scheme the width of the urban highway was determined.[103] Considering a substantial broadening of the highway as one of the measures to be taken, an adaptation of the local zoning scheme would thus be obliged. In the Netherlands, zoning schemes are important instruments in spatial planning. They determine for which purpose real estate can be used and what can be built in specific areas and how one should build. Over the years, environmental standards have become integrated into zoning schemes. Zoning schemes have strict norms for exhaust fumes, noise pollution and levels of soil, water, and air pollution.[104] Adapting the zoning scheme was not considered an option, according to Henk Schroten, at that time general director of Rijkswaterstaat. He did not anticipate the implementation of a final solution in the near future.[105] In this way, the highway's legal embeddedness in a specific zoning scheme contributed to the obduracy of the existing highway design.

Rijkswaterstaat thus proposed not to start with carrying out the suggested temporary measures. Making a general plan and the start of a planning procedure would raise expectations that Rijkswaterstaat would not be able to meet on a short-term basis. In the Second Structural Plan for Traffic and Transportation, the urban highway section of Maastricht was mentioned as a project that would be executed after 2010.[106] Limburg's Rijkswaterstaat Department proposed to start an EIS procedure.[107] This was considered to be less risky than a new zoning scheme procedure on the basis of the trajectory decision of 1982, since a new EIS study would incorporate prevailing laws, measures and opinions.[108]

The Highway's Embeddedness in Legal Regulations

The developments during this phase demonstrate how new and existing judicial regulations may form an important source of contention in processes of urban change. These legal issues also contribute to obduracy. When existing planning structures no longer conform to new judicial regulations, adaptations have to be made and earlier redesign decisions may be nullified. New redesign proposals have to conform to existing regulations as became clear when the Chamber of Commerce proposed its plans for the solution of Liège. One of the reasons for starting a new EIS study was that a better match between recently developed regulations, opinions and laws and new design proposals could be achieved. Furthermore, my discussion suggests how the embeddedness of urban structures in existing judicial regulations, such as zoning schemes or land-use plans, can contribute to their obduracy. In terms of actor-network theorists Callon and Latour, the legal issues formed an "obligatory passage point"[109] in the highway network: entities that were difficult to circumvent or neglect in the process of reshaping the highway's network. Moreover, legal regulations as obligatory passage points represent instances of power because, as we have seen in this case, they play a crucial role in disciplining the interactions of actors.

THE TRAJECTORY/EIS STUDY (1994–1998)

In December 1995, a first basic report outlined the main problems of the urban section of the highway that cuts through Maastricht. The report identified three targets of the Trajectory/EIS Study: to ease traffic circulation (avoid congestion), to enhance the quality of life of those living near the highway (less noise, smells and so forth) and to improve traffic safety. The initial report described the various alternative designs that would be investigated in the study.[110] In 1996, the Rijkswaterstaat Department of Limburg commissioned the national Engineering Department (Bouwdienst) of Rijkswaterstaat to execute a new Trajectory/EIS study. Two formal groups were established in which both the city and Rijkswaterstaat participated: a technical group and a management group.

At this stage, Rijkswaterstaat acknowledged that some of the alternative designs proposed earlier were no longer feasible. Since 1982, all spatial policies of both the city and the Province of Limburg departed from the trajectory decision that had favored the existing trajectory of the highway.[111] The City Board added that residents still widely supported the 1982 solution as well.[112] This preference for the existing trajectory had thus more or less become a given in the city's thinking. The minutes of a City Council meeting showed that all local political parties (still) opposed a diversion east of Maastricht, because the landscape had to be preserved at all cost and a diversion would not necessarily solve all the traffic problems. According to the City Council, the 1982 decision continued to be a sound one and the arguments that were put forward at that time were still considered equally valid, if not more so.[113] Evidently, the city's point of view, in which the ideal of a tunnel and the city border philosophy figured prominently, had undergone no significant changes since the early 1980s. In other words, the existing highway trajectory and the tunnel option had acquired obduracy because of their embeddedness in the 1982 governmental decision, subsequent municipal and provincial policies, and the continued popular support.

Having considered the various forms of embeddedness of Maastricht's highway as an explanation for the difficulties involved in the effort to change the highway's design, I will now discuss and evaluate two major strategies of dealing with embeddedness. Which strategies did the actors employ in their struggle with embedded urban structures? During the phase of the Trajectory/EIS Study two strategies of dealing with the embedded structures of the urban highway can be distinguished: one in which this embeddedness was deliberately, albeit temporarily, disregarded by assuming an almost infinite malleability (the Infralab procedure), and one in which an explicit confrontation with the embedded structures was sought (the Design Workplace). To what extent can these strategies be considered successes?

Disregarding Embeddedness: The Infralab Procedure

Despite the preferences of the city, Rijkswaterstaat tried to broaden the redesign options by restarting the discussion and proposing a number of alternative designs. For example, it once again considered the diversion of the highway along the eastern edge of Maastricht as one of the alternative solutions. The other option was a reconstruction of the existing highway trajectory. Both options could be designed as either a highway (120 kilometers per hour) or a normal road (80 km/h). Besides, an EIS procedure required that the situation of taking no measures at all had to be investigated and the option of keeping the existing situation as it is plus a number of additional measures to reduce congestion. The highway itself could be designed below surface level, at surface level or above surface level.[114] Finally, the "most environmentally friendly" alternative had to be studied as well as the option of a drilled tunnel. These alternatives were all discussed in the first report.[115]

The decision process at this stage was accompanied by a relatively new procedure, based on the "Infralab" methodology[116]. This methodology was introduced by Rijkswaterstaat in 1994 and consists of three steps. First, different users and stakeholders are invited to define the main problems in relation to the infrastructure project involved. Next, stakeholders, users and experts in the field of highway design and civil engineering come together to think about creative solutions to the problems defined in step one. The third and final step involves the selection of the most promising solution by experts and politicians in charge of the issue.

Rijkswaterstaat invited members of groups that had a stake in the urban highway's redesign. Thus, residents, automobile drivers and representatives of the Chamber of Commerce, the city of Maastricht, cyclists' interest groups, Province of Limburg, the General Dutch Automobile Association (ANWB), and local activist groups all met to discuss their view of the main problems and to reflect on solutions.[117] For the ANWB, a good traffic-circulation system without congestion was the most important goal to be achieved, while it also felt that through traffic should not be hindered by local traffic.[118] The group Geen Oosttracé voor de A2! (No East Variant for the A2!), consisting mainly of residents of the eastern districts of Maastricht, wanted to prevent highway construction east of Maastricht. It argued that the quality of life in the eastern part of Maastricht would considerably decrease as a result of a diverted highway while the landscape and ecology would suffer as well. One of its arguments, for instance, was that the east variant might destroy the natural habitat of a rare species of hamster (the "Korenwolf"), which is only found in that part of the Netherlands.[119] Other groups specifically focused on the interests of slow traffic participants, such as pedestrians or cyclists.

Residents of the nearby apartment buildings were also invited. Since 1966, Mrs. O. Kars has been living in one of the apartment buildings near the highway. Mrs. Kars explained that the highway was relatively quiet in the late 1960s and that it was more like a local, two-lane road: "There were no pedestrian crossings. . . . But it was so quiet that I could cross the road easily. No problem at all. During summer heat waves, I slept on the balcony. Now this would be absolutely impossible."[120] Noise pollution was the biggest problem for the residents of the flats: "It is impossible to open a window because it then becomes impossible to have a conversation or to listen to the radio or TV."[121] Despite these problems that gradually aggravated while she has been living near the highway, Mrs. Kars did not want to move: "I do not like to move. It is such a comfortable flat, and it is impossible at the moment to obtain a flat with such a spacious living room and hallway for the same price. . . . Therefore, I stay put, as do the other people who live in these apartment buildings."[122] According to Mrs. Kars, the apartments provide housing to elderly people in particular, many of whom have lived there for a long time and do not want to move anymore.[123]

In 1995, the residents of the apartment buildings near the highway established the Belangengroep Stadstraverse (City Highway Interest Group).[124] According to Mrs. Kars (a member of this group), the residents favor a tunnel with a narrow opening, a local road on top of it, and measures to diminish the quantity of through traffic at the Scharnerweg intersection. Remarkably, some even prefer reconstruction of the existing highway over a diversion. Mrs. Kars explained: "The trajectory is in place, it simply is there. It should be used. And it should be used in a way that is least harmful to the residents."[125]

Interestingly, the Infralab methodology is based on the idea of putting existing solutions and ways of thinking between brackets in order to allow a new creative process to start from scratch.[126] By temporarily disregarding the embeddedness of the urban highway, and by stimulating the generation of ingenuous solutions, the participants were no longer limited by constraints and could thus come up with various radical alternatives. The Infralab method tries to circumvent all the types of obduracy discussed in this book, not only embeddedness. Dominant frames, for instance, related to design solutions that frequently turn up in the process, had to be ignored as much as possible. In this case, the alternatives originally defined in the first report, such as the diversion and the tunnel, had to be set aside temporarily.[127] The fact that users without professional training in highway engineering are involved in the procedure implies that many aspects related to embeddedness and dominant frames (such as the technical criteria for the design of highways and the most appropriate curves) are not automatically taken into account when developing new solutions. According to one of the initiators of the Infralab method, Ad de Rooij, deeply rooted traditions, such as the technocratic approach in policy making and thinking in terms of measures based on strictly scientific reasoning, also need to be bracketed off in the Infralab process.[128]

This approach may produce sweeping proposals for reconfiguration and surprisingly fresh visions. In total 18 new concepts were developed by means of the Infralab procedure in Maastricht. These included, for instance, a tunnel below the Maas River, an elevated road, a new ring road around Maastricht, a combination of rail and road infrastructure in one tunnel, and so forth. As an outcome of the Infralab sessions in Maastricht, a completely new, alternative highway trajectory was taken into consideration: the spiegeltracé (mirror trajectory). This trajectory consisted of a tunnel below the neighborhood of Wittevrouwenveld that formed the mirror image of the existing trajectory. That radically new solutions were proposed in the Infralab process can be partly explained by the fact that the participants did not have to take factors into account that normally cause obduracy. As we have seen in this chapter as well as in the previous one, it can be very difficult during a planning process to set aside dominant frames, existing solutions, laws, policies, personal stakes, assumptions about financial feasibility of solutions or technical design criteria. These factors all play a crucial role in the constitution of obduracy.

Despite the radically new views produced by this procedure, obduracy is hard to bypass. In general it appears to be very difficult for Infralab participants to be genuinely creative. Hans Smeekes, facilitator and co-inventor of the Infralab procedure, explains that this particularly applies to Rijkswaterstaat employees because they know so much and have so much experience that they "must force themselves to abstain from saying too soon: that has already been tried once, that is impossible because. . . ."[129] The Rijkswaterstaat engineers have trouble ignoring their training and expertise in highway design. Infralab employees Hank Kune and Frank van Erkel observe that experts in particular are no longer sensitive to new insights and more radical possibilities for change. As a result, "the interpretation of the process and the solutions still takes place within the "old" frameworks.

Furthermore, participants of creative processes . . . have difficulties to get rid of their own pet subjects. This tends to be the case more often when the group is more homogeneous (more people of Rijkswaterstaat)."[130]

In the course of the Infralab procedure's application in Maastricht, gradually more conditions were set, thus reducing the available elbow room. The experts and politicians that had to choose among the most promising alternatives had to apply four standards to the new solutions: (1) The solutions may not result in an impairment of monuments or buildings of architectural or historical value in the city of Maastricht. (2) The solutions may not result in unacceptable congestion on the major north-south route that cuts through the city. (3) The solution should result in an improved accessibility of the Maastricht region. (4) The solutions should result in an improvement of the quality of life along the highway without transferring problems to another location.[131] These standards, of course, reflect the highway's embeddedness in the larger planning structure and traffic scheme.

A second Infralab procedure was initiated to discuss the most promising outcomes of the first. The aim of the second procedure was to discuss the problems that would be the result of the design solutions chosen in the first Infralab procedure and to find solutions to these problems. [132] The second procedure was thus less "open" than the first. One of the solutions that came out of the first procedure, the spiegeltracé, received specific attention during these sessions. But when this solution was eventually confronted with the existing planning structure it was rejected: the spiegeltracé proved hard to fit into the existing urban configuration. The projected tunnel, for example, could have hardly avoided a nearby cemetery. Most highway engineers and planners consider respecting the integrity of graveyards an unwritten rule. Rijkswaterstaat generally tries to keep a distance of at least 100 meters from existing cemeteries when planning new trajectories.[133] Moreover, the residents of the Wittevrouwenveld neighborhood raised protests against the spiegeltracé. The highway's project manager at Rijkswaterstaat Limburg, Frans Hendrikx, remarked that Wittevrouwenveld used to be a socially disadvantaged area. Much effort had already been put in its upgrading and radical changes might easily disrupt its positive development,

Hendrikx suggested.[134] The spiegeltracé option, then, ran up against a graveyard and a social revitalization program, major factors that turned it into a strongly embedded option after all.

Other design solutions were also rejected during the process when confronted with the existing urban structures. Highway engineers considered the solution of a drilled tunnel as "unrealistic," since this option depended on radical spatial and technological adaptations. Due to the existing highway trajectory's location in an area with high groundwater levels, the upward pressure of the groundwater required a drilled tunnel to be built deep into the soil. This would severely complicate making connections to other roads. In comparison to conventional tunnels, the cost would be enormous. Other solutions were also abandoned for financial reasons. It would be too expensive, for example, to replace or remove underground wiring and cables. Some options were rejected because they would not be able to accommodate future traffic growth, or because they contradicted the goal of improving the quality of life for those living near the highway's trajectory.[135]

Thus, when all the creative potential of the Infralab procedure had been used up and each of the alternatives had been confronted with the commonsense reality of special interest, finances, embedded planning structures and Rijkswaterstaat engineers, the newly invented designs had to be rejected one by one after all. However, this is no reason to argue, I believe, that the Infralab has not been successful. Although the overall procedure rests in part on an imaginary situation of unlimited freedom, it can result in the articulation of new and creative solutions that would not have come up in a process in which users and interest groups are not given a voice. On the other hand, the procedure has a number of obvious disadvantages. First of all, new, creative solutions that are proposed during the Infralab process may lead to protests or even social unrest. The inhabitants of the Wittevrouwenveld neighborhood saw themselves confronted with a solution that, if implemented, would have deeply affected the quality of life in their neighborhood. Second, the process may raise false expectations among the participants. Residents and users may expect that their radical proposals will indeed be implemented, while in most cases this will not be feasible at all. Eventually, the application of the Infralab procedure has to acknowledge the obduracy of urban socio-technology.

Acknowledging Embeddedness: The Design Workplace

The opposite strategy, acknowledging embeddedness in the redesign process from the start, characterizes the Design Workplace approach. A Design Workplace consisting of representatives of Rijkswaterstaat's engineering department, Rijkswaterstaat's regional leadership, the city of Maastricht and a consulting company was established in May 1997.[136] In a series of meetings they tried to think of possible redesigns for the highway that cuts through Maastricht. In this section I will focus on the work of this Design Workplace, which struggled with all the different forms of embeddedness discussed above.

In this episode of the redesign process, tensions between the requirements for the new highway design and the obduracy of the highway that resulted from its embeddedness played a crucial role. Carla Konsten, former project manager of Rijkswaterstaat's engineering department and former chair of the Design Workplace, and Frans Hendrikx, Rijkswaterstaat project manager Highway A2, emphasized the constraining role of rules and legal norms in highway design. The available design options for highways (what is allowed in highway design and what is not) depend on how a particular road is classified. In the Netherlands, guidelines issued by the ROA (Richtlijnen Ontwerp Autosnelwegen) are used to design highways. These guidelines consist of several standardized safety and comfort requirements.

As was pointed out above, the A2 through Maastricht became defined as a "main transport axis" in the Second Structural Plan for Traffic and Transportation.[137] This implied, for example, that designers had to assume a maximum speed of 120 km/h as a guideline.[138] However, in November 1992 Minister of Transportation Hanja Maij-Weggen devalued the urban section of the A2 in Maastricht by re-classifying it as a major secondary road. This decision had consequences for the possible design options. For example, the curve in the existing urban section of the A2 was actually too narrow for a highway, but this problem was solved in the new situation because the road's new classification calls for a maximum speed of only 90 km/h. This lower maximum speed required less space for the curves according to the ROA guidelines, which meant that fewer apartment buildings would have to be demolished.

The history of the effort to redesign the highway is marked by a continuous discussion about the preservation or demolition of elements of the planning structure in which the highway was embedded, in particular the (residential) buildings alongside the highway. The city strongly adhered to the idea of a tunnel, but it also felt that monuments or characteristic buildings near the passage had to be preserved as much as possible.[139] Armand Cremers, a member of the City Board specializing in traffic issues between 1990 and 1998, emphasized the significance of the buildings near the highway in terms of planning. He pointed out that boulevards lined by closed rows of buildings constituted an important pattern in the city's structure. Moreover, the city's eastern part, where the highway is situated, was already more "chaotic" than the western part, and increasing the imbalance would not be advisable.[140] In a 1990 letter to Rijkswaterstaat, the City Board had declared: "We are of the opinion that demolition should be avoided on principle. We therefore insist on the development of a design that actually avoids demolition."[141] To underscore this issue, maps that were made in the context of the redesign effort began to list the Municipal Apartment Building as a "young" landmark, and in its brochures, the city showed pictures of the road in the early 1960s as the desired image for the future: a quiet boulevard lined with old trees and landmark buildings, while the noisy and polluting flow of cars rages on invisibly, in a tunnel, underground.[142]

According to Jan Nakken, a landscape architect and chairman of the Design Workplace, a basic choice had to be made on whether the whole area near the urban highway (including the apartment buildings) should be renewed as a whole, or whether the existing planning structure would be more or less considered as "given," meaning that reconstruction had to take place with as little demolition as possible. The last option was chosen, for the city refused to compromise on the issue of the buildings near the highway.[143] Apart from this argument, costs were also an important consideration: if the option of drastic urban renewal would be chosen, who would have to pay for it? Therefore, the demolition of buildings should be avoided as much as possible, a view that reinforced their obduracy. The Design Workplace gradually defined a speed that required minimal curves and the maximal possibilities for the recovery of the urban fabric as the two major variables in redesigning the highway.[144] At one point, it concluded that only the 90 km/h speed would be a feasible option, because it had the least planning consequences and a minimum of houses needed to be demolished.[145]

The decision to assume a speed of 90 km/h did not solve all problems though. Because the demolition of at least some buildings seemed inevitable, the Design Workplace spent quite some time deciding whether it would be better to do so on the west side or the east side of the highway trajectory.[146] In their deliberations, the role of cultural heritage—monuments or buildings of architectural value—played a crucial role. Although from a planning perspective the public garden on the west side (Koningsplein) was seen as a great asset,[147] eventually the buildings on the east side turned out to be the most obdurate: in case of demolition on the east side, a national monument, the building of the electrical power company of the Province of Limburg, would have to be moved (at an estimated cost of 6 million guilders). Moreover, the apartment buildings on the east side were seen as more valuable than the slightly older buildings on the west side.[148] Despite the loss of the public garden, demolition on the west side involved fewer buildings and fewer monuments and was therefore regarded a better option.[149]

Another important issue that dominated the discussions of the Workplace participants for months involved the transportation of hazardous materials through the projected tunnel.[150] There was a tension between, on the one hand, spatial and environmental concerns and concerns about the quality of life, and design criteria with regard to the safety of tunnels on the other. If hazardous materials were to be transported through the tunnel, its design had to suit that purpose. Official guidelines stipulate that hazardous materials may be transported through an open tunnel, but that they are not allowed in closed tunnels. The chance of accidents involving trucks was a related concern, in particular the danger of smoke accumulation resulting from burning trucks. In the discussions about the design of a closed tunnel, the fire department played an important role, for it had to approve the tunnel's fire safety.[151] This issue is still being regulated regarding tunnels. At the same time, noise standards, defined by law, had to be taken into account as well. Obviously, a more or less open tunnel design would be safer, but it would significantly increase noise pollution.[152]

Rijkswaterstaat had not expressed a preference for a particular design, but the available options were limited by a host of laws and design guidelines, including ROA, noise standards, municipal and provincial zoning schemes, and safety regulations for tunnels. Similarly, the obduracy of concrete elements of the overall planning configuration, such as the buildings on the highway's east side, became more clearly articulated at this stage, and this also limited the available redesign options. Finally, it was concluded in the spring of 1998 that the most promising alternatives were a closed tunnel, a semi-closed tunnel with a narrow opening in the roof, and a ground-floor road with overpass junctions of which the local roads would be lowered. [153]

The Design Workplace's main task involved negotiating the obduracy of (elements of) the urban highway. The participants considered the highway as part of a larger whole—e.g., the overall planning structure, ROA guidelines, and ideas about cultural heritage—rather than as an isolated phenomenon. By thus highlighting the road's embeddedness it became more difficult to come up with radically new alternatives: the participants were aware that changing one element required the adaptation of others, and that a solution that favored certain aspects had negative implications for others. In contrast to the Infralab procedure, citizens were not directly involved in the Design Workplace. It is also important to stress that, like the planner Riek Bakker in the Hoog Catharijne case (chapter 2), most Workplace participants initially were relative outsiders to the long-term effort aimed at reconstructing Maastricht's highway. However, in the end, the solutions proposed by the Design Workplace were not implemented because of changed political and financial priorities at the national governmental level.

In September 1998, during the Trajectory/EIS study, a new government was installed. The new minister involved, Tineke Netelenbos of the PvdA party, introduced a new long-term Program for Infrastructure and Transportation (Meerjarenprogramma voor Infrastructuur en Transport) for the period until 2010. For this entire period, she set aside a total amount of 67 billion guilders, a figure that is not enough to carry out all planned infrastructure projects nationwide. The selection of the numerous projects was based on previous agreements with local authorities, as well as on the extent to which the infrastructure measures were expected to contribute to the Dutch economy at large. Between 1994 and 1998, Minister of Transportation Annemarie Jorritsma had de-emphasized the role of the national government in policies related to local infrastructure projects. This approach had shifted part of the responsibility for traffic problems and congestion from the central government to the local governments.[154] Moreover, at the national level a change of priorities had meanwhile taken place, favoring investments in public transport and shipping. Similarly, in the Province of Limburg priorities had shifted from the south of Limburg to projects in the north.[155] All these factors contributed to Netelenbos's decision not to allocate any funds to the highway project in Maastricht. Instead, she decided to postpone all work on the urban section of the highway through Maastricht until after the year 2012.[156]

Understandably, the city was perplexed. In a letter to the prime minister and the ministers of transportation and housing, the city leadership argued that the minister had failed to take into consideration that the urban highway section not only involved a serious traffic problem, but also a substantial and lingering problem with regard to the quality of urban life. The City Council of Maastricht passed a resolution suggesting that it would be unacceptable to postpone the highway's reconstruction once again (after the 1982 decision). It also demanded that the Trajectory/EIS study be finished. Furthermore, it wanted to find out how the implementation of a plan could be accelerated by considering alternative forms of financing (e.g. a public-private partnership).[157] After deliberations with the minister it was decided that the research already carried out in the context of the Trajectory/EIS study should be completed and that its conclusions

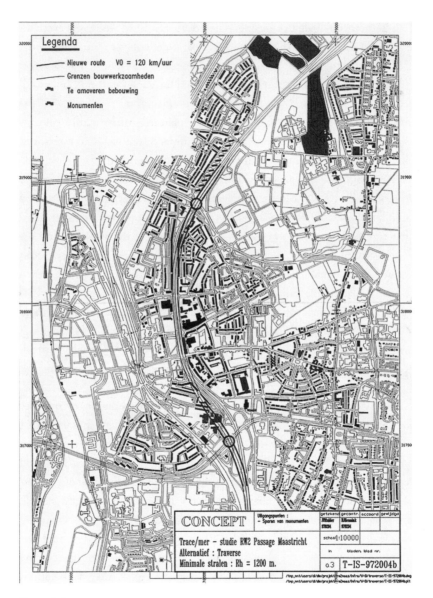

Figure 3.6
Potential new highway trajectory (1997). Source: J. Nakken Zandvoort Ordening & Advies
Utrecht. © Projectbureau A2 Maastricht.

would focus on how to facilitate a public-private partnership that might be willing to fund the highway's redesign.[158]

OBDURACY AND EMBEDDEDNESS

In this chapter, I analyzed the attempts between the 1960s and 1998 to "unbuild" the design of the highway that cuts through Maastricht. Since its initial construction, various more or less radical efforts to adapt the highway have been undertaken. The difficulties involved in implementing new designs could not be explained by referring to material factors alone. It has become clear that several other factors also played a major role, such as ideas about cultural heritage, opinions about Maastricht's overall planning structure, financial considerations, viewpoints about the role of traffic in cities, and legal regulations.

Specifically, I argued that the obduracy of Maastricht's highway could best be explained by focusing on its embeddedness. This theoretical notion refers to the increasing interrelatedness of socio-technical elements as a way to account for urban socio-technology's resistance to change. By stressing the highway's various modes of embeddedness, multiple explanations for the difficulty of changing its design could be advanced and supported. This case shows nicely how the road persists because of the node it occupies in a network of people, laws and regulations, buildings, investments, and so on. In terms of Latour and Callon, it had become an "obligatory passage point" that could not easily be ignored. In the previous chapter it was shown that a high degree of inclusion of actors in specific technological frames resulted in obduracy. Because technological frames are usually tied to specific social groups, this interactionist theoretical perspective only provided a partial explanation of what constitutes obduracy. Major urban structures, however, tend to be embedded in a larger built-up urban environment and this generally causes major challenges when for some reason that structure needs to be redesigned. Obduracy is here explained by its relationship to other actants, rather than by the interests and interpretations of relevant social groups.

The main difficulties involved in the effort to redesign the highway emanated from its embeddedness in the local traffic system, legal regulations, local user practices and the larger planning structure of Maastricht. By focusing my discussion on the relations and interconnections between the various socio-technical elements of the urban section of the highway and the negotiations about the importance of these relations, I demonstrated how, in time, it became increasingly difficult to radically alter the highway's design. The advantage of putting the role of embeddedness in the constitution of obduracy at center stage is that it allows one to better understand the close interrelatedness of the various socio-technical elements that is so characteristic of cities. Embeddedness, however, should not be merely seen as the outcome of a process, because at some stages of what I identified as the "unbuilding" process it may directly influence the urban redesign process. It is precisely this embeddedness that explains *why* specific aspects of a city are difficult to change at a given moment.

It is also important to pay attention to the embeddedness of designs that have not yet been realized. The city and residents of Maastricht appear to favor a tunnel, a preference that gained prominence in particular after 1982, even though as an option it has been around for almost as long as the highway exists. Over the years, this tunnel option became linked to the urban ideals and expectations of residents and local politicians alike. My analysis suggests that designs that still have to be realized may nevertheless become solidly embedded in concrete ideals and expectations as well as in local, regional and national policies, and that, as such, they may influence the unbuilding process for long periods of time.

The difficulties involved in dealing with various forms of embeddedness at the same time stood out most clearly in my discussion of the effort of the Design Workplace. Its effort was directed at dealing with the embeddedness of the highway rather than overcoming dominant frames or traditions. Hence, the design options it generated were much more constrained than the Infralab results. In the Infralab method, the various forms of obduracy discussed in this book were deliberately, albeit temporarily, disregarded. This enabled the articulation of radically new design solutions.

Despite all efforts, no solution to the problems around Maastricht's highway has been implemented so far. In 1999 a new traffic lights regime was installed at the highway section through Maastricht. Some people believed that the problems were solved by that simple measure. But Rijkswaterstaat and the City Board of Maastricht emphasized that the ultimate solution could only be a tunnel. Much effort has been put in lobbying in Den Haag, and at the moment, the City Board of Maastricht is quite optimistic about the chances that the building activities for a tunnel will actually start in 2007.

THE SPATIAL RENEWAL OF THE BIJLMERMEER

The Bijlmermeer, a suburban district of Amsterdam, was built in the 1960s and the 1970s. Based on design principles that originated in the modernist planning tradition, it had concrete high-rise apartment buildings, large open public spaces, a separation of traffic flows, and an orthogonal layout. The planners of the Bijlmermeer were quite optimistic about their "City of Tomorrow," which catered to middle-class families in particular. The many shared facilities were meant to strengthen the sense of community among residents. Moreover, the overall design's uniformity, deliberately not expressing differences in individual life styles, reflected the designers' ideal of social equality.

In the 1970s and the 1980s, however, the Bijlmermeer became one of the most harshly criticized suburban projects of the Netherlands. It elicited much controversy, and understandably so, for the noble social ideals of the designers and the social dynamic that actually evolved could not have been more at odds. It soon became clear that the neighborhood had little appeal, in part because of its then fairly isolated location on the city's edge, and quite rapidly it deteriorated into Amsterdam's most impoverished and crime-ridden suburban district. After years of contention and ardent debates, in which community workers, Bijlmermeer residents, local politicians, neighborhood council members, architects, planners, and housing officials took part, it was decided in the early 1990s that in order to solve the problems a rigorous spatial renewal would be necessary, including the demolition of 25 percent of the high-rise apartment buildings. In 1999 the authorities decided in favor of an even more rigorous renewal of the Bijlmermeer: to demolish another half of the apartment buildings and replace them with low-rise areas.

Several other modernist neighborhoods over the world suffered the same fate as the Bijlmermeer. The history of the Pruitt-Igoe housing project in St. Louis has many similarities with the Bijlmermeer. The Pruitt-Igoe complex was built in 1954–1956. Its demolition in the early 1970s had an enormous impact on the debate about public housing in the United States. After the 1970s, public housing became harshly criticized in the United States. In the United States as well as in the Netherlands, postwar public housing was populated by socially weak groups, consisting of tenants with low incomes. The housing complexes were large scale and impersonal and the exploitation of these houses was non-profitable. In both the Bijlmermeer and Pruitt-Igoe, a high turnover of apartments and the fact that the buildings generally did not comply to the living preferences of most people, formed important aspects of the problem.[1] According to the planning professor Peter Hall, the demolition of Pruitt-Igoe, which was covered "live" on TV, became "an instant symbol of all that was perceived wrong with urban renewal, not merely in the United Stated but in the world at large."[2]

Another housing project in the United Kingdom, the Hutchesontown Gorbals in Glasgow, faced similar problems. These buildings were built in the period 1960–1966 and were demolished in September 1993. Like the Bijlmermeer and the Pruitt-Igoe complex, these high-rise complexes were built by architects who were inspired by modernism. The context in which these blocks were built was also very much comparable to the Bijlmermeer: housing was in short supply and economic conditions were difficult.[3] Several reasons have been mentioned for the failure of Hutchesontown, many of which are also applicable to the Bijlmermeer: the complex suffered from high maintenance costs and bad maintenance (electricity failed, water penetrated, windows were decaying), there was no on-site management in the form of a concierge, and nobody cared who lived in the Hutchesontown block and whether they "could afford to pay the electricity bills."[4]

Brasília, the well-known city in Brazil that was completely built according to functionalist ideals, did not live up to the expectations of its residents and designers either. Similar problems occurred as in the Bijlmermeer. People felt that the city was too anonymous and that the huge apartment buildings were too much standardized. Like the design of the Bijlmermeer, all apartment blocks were similar from the exterior. Residents felt disoriented because everything looked the same—a problem that is also frequently mentioned in critiques on the Bijlmermeer. And like the Bijlmermeer, Brasília had the problem that, despite the wish to construct an egalitarian society by homogenous housing types, the contrary effect was reached: the poor people lived on the outskirts of the city, whereas the rich elite lived close to the downtown area.[5] Of course, the main difference with the present situation in the Bijlmermeer is that in Brasília no serious attempts have been made so far to radically change the urban structure.

PLANNING AND BUILDING THE BIJLMERMEER

The Bijlmermeer, designed and built in the 1960s and the early 1970s,[6] was planned as part of Amsterdam's General Extension Plan by the city's Planning Department of Public Works. The design of the Bijlmermeer relied on a specific interpretation of the modernist tradition in architecture, its main features being the strict separation of traffic types and the homogeneity of building types. The Bijlmermeer was conceived as a decidedly functionalist urban district, assuming a stringent separation of living and working, of traffic and recreation. The architect Rem Koolhaas argued that "the Bijlmermeer is the product of an architectural dogma—most considerately practiced by CIAM in the 1930s—but realized with retrospective effect."[7]

Le Corbusier (1887–1965) is generally regarded as the founding father of the principle of functional separation. His ideas were adopted by CIAM (Congrès Internationaux d'Architecture Moderne), an international group of modernist architects who tried to anticipate the challenges that societal changes and the increasing world population were creating for industrialization, housing, and planning.[8] The sociologist Magali Sarfatti Larson, who studied the shifts in twentieth-century architectural discourses, points out that in this area the first and most far-reaching change occurred in the 1920s with the emergence of the Modern Movement, in which theorists and architects from Germany and the Netherlands participated in particular. After World War II, an adapted version of European modernism became known as the International Architectural Style.[9] In this postwar phase, modernism became also influential in the United States, and, according to the architecture professor Diane Ghirardo, modernism became associated with the power of capitalism. Apart from architects and planners, urban developers became proponents of modernist buildings as well because they were cheaper and much quicker to build than traditional building styles. The sociologist David Brain hints at a distinction that is often made between European and American modernism in architecture: in the United States, modernism would be more pragmatically and less ideologically motivated than in Europe.[10] However, in the 1960s, the fiercest attacks on modernism (e.g. by Jane Jacobs and Lewis Mumford) also originated in the United States.[11]

CIAM was established in 1928 in La Sarraz. The main principle of CIAM was that planning should be based on functional considerations rather than aesthetic ones. "The Functional City" was the main topic at the fourth CIAM conference, held in 1933 in Athens. The conference's chair, Cor van Eesteren, chief architect of the Urban Development Section of Amsterdam's Public Works Department since 1929, was one of the people in charge of the city's General Extension Plan. Based on an analysis of major urban problems, CIAM architects formulated a number of basic analytical guidelines for the design of new cities: housing should not be situated too close to roads for traffic, high-rise buildings had the advantage of allowing more space for recreational purposes, green open spaces should be distributed equally throughout the city, residential and commercial areas should be not too far apart and they should be well connected, and the separation of the various flows of traffic can be realized by creating overpass junctions. Le Corbusier reformulated these principles in a document that became known as the Charter of Athens; it is viewed as the doctrine of modernist planning.[12] Based on these assumptions, Le Corbusier designed a number of plans, including "Plan Voisin" (1925) and "La Ville Radieuse" (1930), that became major sources of inspiration for planners and architects.[13]

"The contribution of Le Corbusier's utopian city to the twentieth century can be seen in the tower-block housing estates which went up on the outskirts of all the major Western cities," Pamela Neville-Sington and David Sington claim in their book on the influence of utopianism on the modern world.[14] However, the historian of architecture Auke Van der Woud argues that in postwar planning CIAM has been less influential than is usually assumed. Although many urban districts that were built in the postwar era suggest some sort of functional separation, like, for instance, those consisting of high-rise apartment blocks in rows surrounded by green public spaces, it is not always correct to label them as instances of CIAM planning. Van der Woud conjectures that the views of CIAM were spread via rather "obscure" magazines, which renders it doubtful whether those in charge of planning new neighborhoods were actually familiar with them. Moreover, the rationalization of the construction business after World War II was a consequence of concrete postwar circumstances, such as the great lack of housing and the rapid population growth, rather than a "logical" effect of CIAM. Van der Woud also suggests that CIAM proponents have put much effort in claiming that their viewpoints were at the basis of these planning developments. This "optical illusion," he claims, "came into being because postwar housing was built in accordance with the same methods as those which had been analyzed and propagated in the twenties by the then modern architects and this created related architectural images, especially in the sixties, the period of widespread high-rise building."[15]

The architecture of the Modern Movement, Le Corbusier's architecture in particular, has often been situated in a longer-term utopian tradition of planning. The philosopher of technology Hans Achterhuis distinguishes three characteristics of utopia: (1) Utopia assumes a strong sense of constructability and controllability. (2) It is aimed at the establishment of an (ideal) community. (3) It relies on totalitarianism to achieve radical social change.[16] These characteristics can also be applied to the work of Le Corbusier. Neville-Sington and Sington make clear that Le Corbusier saw architecture as a means of control and the city as a machine.[17] The city could contribute to making life orderly, balanced and efficient.[18] Suggesting that Le Corbusier's architecture was aimed at changing social behavior "without the need for outright political revolution,"[19] the two authors situate this aim in a longer and influential tradition of "salvation by bricks alone": "The ideal city was not just the location of the perfect society, it was the means to bring it about."[20]

With regard to utopianism in planning, a slightly different interpretation can be given of Achterhuis's third characteristic of utopia. Totalitarianism can also be understood as the sense of totality or rigorous consistency that is usually strived for in the modernist tradition. This becomes particularly clear in the widespread modernist practice of designing blueprints or "master plans" that imagine altogether new worlds—altogether new, "ideal" communities, cities or societies. The features attributed to modernist architecture by Diane Ghirardo correspond very well with this analysis. Ghirardo emphasizes that among the most important characteristics of modernism in architecture are its technological determinism, the belief in the power of architecture to transform the world, and the idea that social problems can be solved through a rationalist approach. Moreover, Ghirardo observes a relative "indifference" to history and tradition among modernist architects.[21] James Scott emphasizes the totalitarian features of Le Corbusier's thinking: "Le Corbusier sees himself as a technical genius and demands power in the name of his truths. Technocracy, in his instance, is the belief that the human problem of urban design has a unique solution, which an expert can discover and execute. Deciding such technical matters by politics and bargaining would lead to the wrong solution."[22]

A number of political and social values that can be characterized as utopian were also at the basis of the Bijlmermeer design.[23] One of the challenges for the project's designers in the 1960s was to produce a new housing concept, a new, suburban type of living for a specific group of citizens: middle-class families owning a car.[24] The designers clearly considered social equality a major principle or value, since the plan left little or no room for expressing a sense of individuality or an individualist lifestyle; not surprisingly, the design of the

Bijlmermeer has often been characterized as egalitarian or even socialist.[25] Another value that the design was meant to promote was a sense of community: the various apartments and car parks could be reached by means of indoor walkways, along which meeting places, shops, leisure rooms and day nurseries were to be situated.[26] At the same time, everyone's privacy had to be protected, an issue that was seen as a sine qua non for the growth of any happy community. The homes and apartments were designed to create "as much privacy as possible for the family inside their house," but outside "the mutual contact between the residents" had to be optimally stimulated by the design as well.[27]

Scaling up was another aim of the designers and architects of the Bijlmermeer. In their view, increasing urbanization would inevitably result in a more functional spatial division between work, housing, mobility and all sorts of facilities. To create uninterrupted areas of public green space and to guarantee maximum privacy, huge blocks of high-rise apartment buildings were planned. These were constructed by relying on the comparatively inexpensive and fast method of system building. In the Bijlmermeer, this resulted in a design based on nine-story apartment buildings shaped in the form of honeycomb ensembles (figure 4.1).[28] Ninety percent of all housing in the Bijlmermeer consisted of these high-rise buildings, while the remaining 10 percent were projected as single-family homes. Although the Bijlmermeer planning team hotly debated the exact number of high-rise buildings, the housing shortage and the ensuing pressure to build new homes caused it to opt for the fast and effective solution of high-rise apartment buildings.[29]

In the 1960s, there was great concern in the Netherlands for future lack of space as a result of the expected growth of the population. During World War II, many homes were destroyed and the construction of houses virtually stopped.[30] In the decades immediately after the war, housing shortage was high on the political agenda. In 1968, Das, Leeflang, and Rothuizen proposed several technical measures to intensify the use of land set aside for urban growth.[31] Instead of "space-consuming" low-rise buildings, they expressed a preference for huge urban residential towers. They referred to the Bijlmermeer as an example of a project approximating the ideal of concentrated urbanization, but they felt

Figure 4.1
Aerial view of typical Bijlmer apartment buildings (1971). Source: Archive Dienst Wonen
Amsterdam. © Dienst Wonen Amsterdam.

that the planned high-rises and multi-story car parks were not radical enough, suggesting that space was still wasted.

Aside from the apartment buildings, the traffic system served as an important structuring element in the Bijlmermeer project. According to the designers, the area's infrastructure of roads had to provide a solid basis for its future design: "Traffic and transportation systems and the design of city neighborhoods constitute an indissoluble unity, both from a functional point of view and (partly because of that) from a formal point of view. . . . A major divergence from the configuration's basic layout and assumptions is impossible without being detrimental to its functionality."[32]

One of the ideals on the minds of the planners involved the creation of a living space, a spacious green zone that would not be accessible to cars.[33] The residents should feel like they were living in a park. In their effort to achieve this effect, the planners and architects of the Bijlmermeer tried to avoid some of the "errors" that had been made in the Amsterdam garden cities that were built in the 1950s.[34] Concretely, this resulted in banning car traffic from the immediate living environment of the residents.[35] Automobiles were exclusively allowed on the semi-elevated roads that gave access to the large multi-story car parks. The size of the car parks was based on the expectation of increasing car ownership, projecting 1.5 cars per family. Pedestrians were not allowed on the semi-elevated roads. For them, the planners created an autonomous system of ground-level walking routes. (See figures 4.2 and 4.3.)

Although some of the design principles were already abandoned before the project's completion, the Bijlmermeer is generally perceived as a far-reaching application of modernist planning, infused with socialist and egalitarian principles. During its construction, severe financial problems occurred, which inevitably led to budget cuts.[36] Elevators, telephone boxes, waiting rooms, shops, and mailboxes were planned but never fully realized. Furthermore, the building of the multi-functional car parks led to delays in the planning process, which in turn delayed the building of the shopping centers that were situated below the indoor parking facilities.[37] Moreover, in the early 1970s the network of bicycle paths, roads, and public transportation (the subway) failed to function properly.[38]

Figure 4.2
A part of the Bijlmerdreef from which pedestrians and cyclists are prohibited (1998). Photograph by A. Hommels.

Figure 4.3
Separate roads for pedestrians and cyclists (1998). Photograph by A. Hommels.

Thus, a number of modernist planning principles became embedded in the Bijlmermeer's overall configuration, including the apartment buildings, road system, indoor walkways, car parks and open public spaces. Utopian plans like Le Corbusier's have been blamed for their failure to take into consideration "the way real flesh-and-blood people live and act."[39] In fact, architects tend to be frequently criticized for their unrealistic views of society, if they are not characterized as "poor sociologists," as one Dutch journalist does in an article about the Bijlmermeer.[40] Many times utopian urban ideals are only partially realized in concrete modernist plans, and this is not without consequences for our urban environments: "Real cities do not remain as static as the ideal cities must to retain its special status. . . . Therefore the schemes of men like . . . Le Corbusier can never be fully realized; yet their partial implementation has had a tremendous impact on our landscape. We live surrounded by the fragments of their utopian dreams."[41]

Indeed, once the neighborhood was almost finished, the Bijlmermeer failed to live up to the expectations of the planners and the inhabitants. Quite soon, the Bijlmermeer became heavily criticized on different levels. Residents criticized the partial execution of the plans, the finances and management and the initial absence of residential facilities. Moreover, despite the initially enthusiastic reactions to the Bijlmermeer concept,[42] later on its planning concept was heavily criticized as being "superseded" by social developments.[43] Already in the 1960s, the profession of planning firmly condemned the uniformity and monotony of town districts.[44] Once an icon of modernist architecture, the Bijlmermeer quickly became a symbol of the *failure* of modernist architecture. Functionalist views were often seen as the principle cause of what was called the "dehumanization of modernist town planning."[45] The authors of a study of the Bijlmermeer suggested that its planning concept "became more and more an oppressive predicament" that proved hard to reverse.[46] The Bijlmermeer's modernist, functionalist ideals were seen as outdated and the neighborhood became viewed as an "anachronism."[47] In another report the Bijlmermeer was characterized as poorly planned, especially because of its overall inflexibility; the authors felt that to most residents the project was largely "dysfunctional."[48]

Some critics perceived a fundamental discrepancy between the practical aspects of the Bijlmermeer's design and the socio-cultural developments in Dutch society during the closing decades of the twentieth century, and they believe that this is the main reason for its "failure." As a residential concept, the Bijlmermeer did not cater to a particular social need. "One reason for the technical failure of the Bijlmermeer," the planner Dirk Frieling suggested, "is that the technical concept has never had an administrative form to match, and it seems as though, even now, no one is really interested in finding one."[49] And Bernadette de Wit, a journalist and a Bijlmermeer resident, argued that "the institutions had no answer to the specific demands of a neighborhood like the Bijlmermeer; their traditional routines and procedures did not fit our whimsical reality."[50]

Besides this critique of its design and architecture, the public image of the Bijlmermeer became very negative for other reasons. Plagued by unemployment and crime, the Bijlmermeer had become one of the most criticized urban districts of the Netherlands by the 1980s. It became associated with crime, drug use, unemployment, disorder, vandalism, illegal immigrants and unattractive architecture. In time, the public image of the Bijlmermeer became so negative that Bijlmermeer residents felt the responsibility to defend their neighborhood by publishing a report characterizing the Bijlmermeer as "in many respects a failure of a number of administrators, town planners, planners and other experts" but asserting that "the Bijlmermeer is certainly not the hopeless case it is often portrayed to be in both the professional and tabloid press since the first residents moved in" and that "the inhabitants of the Bijlmermeer wish to fight this image of the neighborhood as a cesspit of misery, crime, vandalism and vice."[51] De Wit argued that cynicism about the Bijlmermeer increased further after the mid 1980s. Politicians no longer believed in it: "They abided with the half-finished product they ended up with and no longer tried to complete it with respect for the original design."[52]

Below, I will analyze the various attempts of the actors involved—Bijlmermeer residents, policy makers, planners, architects, and community workers—to either change or preserve the Bijlmermeer's original concept. Many solutions have been put forward since the mid 1970s, including the more recent proposals to demolish parts of the Bijlmermeer and to formulate an altogether new planning structure. In describing these endeavors, I make a rough distinction between three phases. The first phase comprises the years between 1974 and 1986, during which only minor adaptations were proposed while the existing planning structure of the Bijlmermeer was largely preserved. In the second phase, between 1986 and 1992, the basic outlook of the Bijlmermeer and its planning principles were no longer taken for granted, which prompted the first concrete proposals for demolition and fundamental redesign. In the third phase, from 1992 up to the present, the spatial renewal actually got underway.

MAINTAINING AND IMPROVING THE ESTABLISHED PLANNING STRUCTURE (1974–1986)

From the outset it had been clear that the ways in which the Bijlmermeer residents used and experienced their new environment did not always correspond well with the intentions of the designers of the Bijlmermeer. As Lucas van Herwaarden, a Bijlmermeer resident and a local landscape architect, observed, "the first cracks in the ideal image invented by the specialists were induced by usage."[53] One illustration is that residents started to use the huge, empty, freely accessible multi-story car parks for commercial activities, such as restaurants and shops, while homeless people and drug addicts soon discovered them as good places to spend the night. Furthermore, instead of appreciating the large areas of shared public space, Bijlmermeer residents began to divide some of them into small private lots for gardening. The originally planned public footpaths were diverted by frequent use, and balconies and other public spaces were used for the dumping of garbage. These interpretations and practices indicated a break with some of the ideas originally embedded in the Bijlmermeer concept. In this way, the actual usage of the built environment already contributed to its "unbuilding." However, other, more formal attempts at adapting the Bijlmermeer to the needs of its residents appeared to be more difficult to execute.

In 1974, the Housing Department of Amsterdam established the Bijlmermeer Management Group, to be directed by the architect and Bijlmermeer resident Pi de Bruijn. This group tried to improve the Bijlmermeer's living environment by proposing a range of practical measures. De Bruijn was very frustrated, for example, by what may seem an inconspicuous detail: the great number of doorsteps in the Bijlmermeer apartment buildings. He lived on the eighth floor, and when he needed to go and get something from his storage space in the basement he had to cross twelve doorsteps, each of which he saw as a disturbing obstacle. To gain support for this concern, he established a group called Bijlmer Drempelvrij (Bijlmer Without Doorsteps).[54] Another initiative of the Bijlmermeer Management Group involved its proposal to set aside parts of the large areas of green public space for smaller private gardens. In de Bruijn's view, the Bijlmermeer's design failed to cater to some of the practical everyday needs of residents: ". . . there are many places that are not accessible by cars. . . . There is an ill-defined and often very messy area between the public roads and the front doors of the apartment buildings. Imagine that you are going to die in one of those apartments: how can you get out of it? . . . Or suppose that you want to get married—a social event that most view as a special occasion. I noticed that [the planners] had not thought about that. . . ."[55]

De Bruijn explained that some minor adaptations were made to improve these situations, but that other changes appeared to be very difficult to realize. The planners and architects who designed the Bijlmermeer were against all proposals that did not comply with their view of the Bijlmermeer as a modernist town district based on a socialist-egalitarian concept. Moreover, the Planning Department legally consolidated the existing layout in a zoning plan that was ratified in 1974.[56] As a consequence, de Bruijn contended, it was impossible to put in extra windows in a number of apartments or to create new apartments at the ground-floor level, because for some reason this did not comply with the overall design's egalitarian principle.[57] He left the group in 1977 because he felt the problems had grown too complex and wide-ranging.

In 1980, a group of Bijlmermeer residents joined forces in a working group, called the Bijlmermeer Community Workers Foundation, and their aim was to write a report based on their experiences. They sought to make a positive contribution to the solving of problems in their neighborhood. The problems they identified ranged from larger social issues, such as the "imbalanced makeup of the Bijlmermeer population," and issues related to the overall design, such as noisy apartments, bad signposting, and long walking distances, to quite concrete problems associated with the availability of facilities and their maintenance. A solution to each of the problems was formulated.[58]

Before the mid 1980s, policy makers defined the central problem of the Bijlmermeer mainly as a housing problem. In 1982, the city and the Amsterdam Federation of Housing Corporations founded the Project Group on High-Rise Buildings in the Bijlmermeer.[59] In a report titled "A Plan for the Bijlmer," they limited their concern to housing problems that could be solved in an instrumental manner.[60] According to the authors, the central problem was the increasing number of vacancies in the high-rise apartment buildings of the Bijlmermeer.[61] Already in 1970, the Housing Department of Amsterdam noted in a report that the Bijlmermeer's average percentage of rented apartments and homes was not as high as expected. A 1983 report of the Amsterdam Council for Urban Development concluded that this problem was caused by a number of unforeseeable developments: (1) The Bijlmermeer's population turned out to be quite different from what had been anticipated: while members of the targeted middle class proved to be only marginally interested in living in the Bijlmermeer, various new groups of migrants and workers from other countries did end up there. (2) Many planned social and cultural facilities and shopping centers had never been completed. (3) In the economically constrained years many of the residents were seeking cheaper accommodation because of the high rents and service costs in the Bijlmermeer.[62] The authors concluded, in short, that all the facilities that had been planned should be completed.

In the 1970s, the Bijlmermeer evolved into a haven for members of various marginal social groups, such as refugees, migrants, and illegal immigrants. Moreover, the drug scene that was previously based in parts of downtown Amsterdam had moved to the Bijlmermeer. As a result, drug use and crime increased in the Bijlmermeer and its public image became increasingly negative.[63] After Surinam, a former Dutch colony, gained independence in 1975, many Surinamese people, who were given the opportunity to remain Dutch citizens, moved to the Netherlands and ended up in the Bijlmermeer.[64] Partly as a result of Amsterdam's malfunctioning system of housing distribution, some of the apartments became overcrowded. One of the apartment buildings, named Gliphoeve, was particularly notorious in those days. The Amsterdam Federation of Housing Corporations concluded that this building had to be either demolished or radically changed. At that time, however, in the eyes of both the City Board and the national government demolition was out of the question: the buildings were only eight years old and there had always been a housing shortage in Amsterdam.[65] It was therefore decided that the building ought to be completely remodeled, involving, among other things, the closing off of the indoor walkway and the realization of a new road where residents could park their cars adjacent to the building.[66]

The Project Group on High-Rise Buildings in the Bijlmermeer concluded that these buildings did not attract many people. It explicitly disputed the opinion that the Bijlmermeer planning concept might be substandard and proposed a new policy for the Bijlmermeer to achieve differentiation in the high-rise areas. On purpose, though, this policy excluded the option of a radical change of the original planning principles. It emphasized that the planning principles of the Bijlmermeer should not be relinquished, not even regarding details.[67] The Project Group also proposed a number of practical measures such as lowering rents, closing off interior walkways, adding elevators, providing ground-level parking areas, adding ground-floor apartments, improving promotional activities, fighting drugs, and so forth. Furthermore, it proposed to replace the various existing housing corporations by a single new one, which would be in charge of rentals and management for the entire Bijlmermeer. In 1984 a new housing corporation, called Nieuw Amsterdam, was established.

Explaining Obduracy: The Role of Persistent Traditions

The efforts to either change or preserve the Bijlmermeer in the period between the mid 1970s and the mid 1980s can be summed up as follows. First of all, the local practices of Bijlmermeer residents who started to make different use of the Bijlmermeer contributed to a softening of the Bijlmermeer concept. In official attempts to adapt the Bijlmermeer, the idea that the high-rise buildings should be preserved in their existing form, as well as the notion that management of the apartment buildings and the living environment should be intensified, was maintained as a policy guideline.[68] Although the Amsterdam Federation for Housing Corporations proposed demolition in the case of the Gliphoeve building, this never was truly a serious option. In this period, the proposals assume either the "finished" character of the neighborhood or the idea that the Bijlmermeer is still "unfinished" and ought to be completed along the lines of the original concept. Although some of the adjustments were only minor, such as closing off interior walkways or removing thresholds, other measures, such as experiments with parking at street level and the idea of creating more differentiated apartment buildings, meant a more significant deviation from the original concept. In general, however, measures were not aimed at a radical deviation from the original planning principles. According to de Bruijn, chair of the Bijlmermeer Management Group, the Planning Department blocked all of the more radical attempts at change proposed by Bijlmermeer residents and the Bijlmermeer Management Group (such as creating extra windows in a number of apartments or to create some apartments on the ground floor) that did not comply with the department's view of the Bijlmermeer.[69]

In this phase the design of Bijlmermeer turns out to be relatively obdurate. How can this obduracy be explained? In contrast to the case of Hoog Catharijne, opposing groups with rigid technological frames do not dominate the discussions here. Likewise, in this stage the embeddedness of Bijlmermeer structures does not appear to be a proper explanation for the difficulty to change the Bijlmermeer. I would argue that the main reason why the Bijlmermeer was difficult to change, was because of the wish to adhere to the original ideas behind the Bijlmermeer design. This first stage of rethinking the Bijlmermeer shows a striving for conceptual consistency of planners, architects, and politicians involved in the planning of the Bijlmermeer: changes were allowed as long as they were in line with their view of the Bijlmermeer's overall concept. In their arguments to preserve the Bijlmermeer, policy makers and architects explicitly mobilized the planning tradition of the Bijlmermeer. During this phase, deviation from this tradition was generally not considered the right way to solve the Bijlmermeer problems. According to de Bruijn, one aspect of the Bijlmermeer tradition was particularly emphasized in this phase: its socialist-egalitarian roots. Thus, the relative obduracy of the Bijlmermeer in this phase can be at least partly explained by the importance attached to the planning tradition on which the Bijlmermeer design was based. In theoretical terms, we may conclude that the Bijlmermeer tradition had such a strong momentum, that a deviation from this "path" was not seen as a viable option in those days.

However, according to René Grotendorst, employee of the Amsterdam Federation of Housing Corporations between 1980 and 1984, other reasons for not proposing radical modifications also played a role. For one thing, the buildings were still fairly new, so demolition was not considered to be a reasonable option. There had been a housing shortage in Amsterdam for years and this fact also sharply conflicted with the idea of demolishing buildings. Moreover, most agreed on the quality of the Bijlmermeer apartment buildings and that they were better than those in most other districts of Amsterdam. It made perfect sense, therefore, to spare the Bijlmermeer apartment buildings.[70]

THE BIJLMERMEER: NO LONGER TAKEN FOR GRANTED (1986–1992)

The measures of the early to mid 1980s had concentrated on streamlining the management of the Bijlmermeer (by establishing one housing corporation), on upgrading the ground-floor level of the apartment buildings, on improving the safety and quality of the apartments, while the rents were lowered as well. These changes were meant to substantially reduce the fairly high number of unoccupied apartments in the Bijlmermeer. By 1986 it had become clear that these measures failed to work, for as much as a quarter of the total number of apartments was not rented out. Moreover, crime and drug use increased further.

These negative developments signaled that a more radical approach had to be considered. In 1986, the housing corporation, the city government and the national government established a working group called Toekomst Bijlmermeer (Future of the Bijlmermeer). In a report, Toekomst Bijlmermeer presented five scenarios, four of which proposed replacing some of the large apartment buildings with new housing. According to ter Horst et al., the report's major concern was "that the improvement of the Bijlmermeer could only be achieved by changing the housing supply."[71] Although none of the four scenarios that involved demolition of buildings acquired the support of the parties—the city, the government and the housing corporation preferred the fifth scenario[72]—demolition nevertheless became a hotly debated issue from that point on.

In 1987, Amsterdam's Housing Department argued that demolition was not feasible.[73] Its view was based on the following reasoning: demolition found no widespread support among the population, it involved too much a "technical" approach, it would seriously restrict alternative options for improvement, it meant capital destruction and, finally, it ignored results that had already been achieved as well as present planning options. The Housing Department concluded that the present role of the Bijlmermeer on the housing market should be accepted. It criticized the past approach in which the Bijlmermeer problems were defined as limited to housing problems, and emphasized that there was a social dimension to the problems as well. Therefore, it recommended more monitoring and intensive management, while demolition should only be reconsidered if no other solution could be found at all.[74]

This proposal was sent to Deputy Minister of Housing Enneüs Heerma, who rejected it in September 1988. Unwilling to accept the deteriorating situation of the Bijlmermeer and favoring a more structural approach, he proposed to establish a new commission, again called Werkgroep Toekomst Bijlmermeer (Working Group Future Bijlmermeer).[75] In the meantime, the financial position of Nieuw Amsterdam, the housing corporation in charge of the Bijlmermeer, had seriously weakened because of the high management costs. Moreover, the occupancy rate of the apartment buildings had worsened rather than improved, thus further undermining the financial position of Nieuw Amsterdam. Understandably, then, the new working group's primary task was to explore how this trend could be reversed. Its report, titled De Bijlmermeer blijft, veranderen (The Bijlmermeer Remains, Open to Change), was published in 1990. In this report, the option of demolition cropped up again, quite dramatically, in a reference to a report by the consulting agency Kolpron that included interviews with potential private investors. In order to transform the Bijlmermeer into an attractive residential area in the future, these investors argued that radical changes of the existing planning structure would be inevitable. Especially the "homogeneous population structure" had to be modified. The investors concluded that a maximum potential could only be guaranteed by demolishing the whole neighborhood.[76]

The working group, however, did not share this view. It sketched two scenarios for the future of the Bijlmermeer. One scenario was based on the present planning structure in which a number of instruments were recommended to improve the housing corporation's financial situation.[77] Another scenario delineated a radical change of the present planning structure. Specifically, it proposed the demolition of 25 percent of the high-rise apartment buildings, to be replaced with new low-rise housing; the upgrading of a number of the buildings to a more expensive market segment; a basic remodeling of the remaining apartment buildings; and making improvements in the infrastructure. The working group, in fact, favored a combination of the two scenarios, but it saw the original planning structure as the root of all of the problems. Its report therefore underlined the necessity of demolition by suggesting that scenarios that would hold onto the old planning structure might succeed in temporarily alleviating the financial problems of the housing corporation, but they would fail to address the "structural problems of the Bijlmermeer."[78]

Gradually, more voices were raised against preserving the original layout of the Bijlmermeer, its high-rise apartment buildings in particular. According to architects from the Office for Metropolitan Architecture (a private business), demolition could become a serious option because the public image of the Bijlmermeer had developed into a very negative one.[79] In part because of its high crime rate, it was perceived as one of the most troubled neighborhoods in the country, and this nationwide attention forced local policy makers, politicians and the managing agency to question the (once) obdurate planning structures of the Bijlmermeer. Moreover, the city leadership of Amsterdam did no longer want to be responsible for the financial debts of the managing housing corporation.[80] At that point, it could no longer be denied that radical intervention was called for.

Although the authorities acknowledged the Bijlmermeer's social problems, these were not the principle reason for the fundamental changes that would be proposed. According to Tineke van den Klinkenberg, a member of the Renewal Bijlmermeer Steering Committee (see below), the decision to build the Amsterdam Arena next to the Bijlmermeer also played a crucial role in the decision to renew the Bijlmermeer. It hardly seemed reasonable to build a hypermodern stadium close to a deteriorated town district and ignore its problems.[81] Gradually, a consensus emerged on the need for a fundamental adaptation of its planning structure. Only if its original concept would be given up at least in part, the Bijlmermeer could be given a second life. The "demolition of apartments" was seen as "one of the instruments" that needed to be deployed to effect a break with the past.[82]

In November 1990, the city finally agreed with the plans put forward by the working group. The managing housing corporation also favored demolition of apartment buildings since it felt that its own survival, as well as that of the Bijlmermeer as a whole, depended on such radical intervention. Although in 1986 the corporation had voted against demolition, now it was prepared to accept the plan because of its poor financial situation.[83]

In 1991, the three partners (the city of Amsterdam, the neighborhood board of South-East,[84] and Nieuw Amsterdam) established the Renewal Bijlmermeer Steering Committee.[85] The planner Dirk Frieling was put in charge of the committee, which consisted of representatives from Nieuw Amsterdam, the neighborhood board, and the city and two external advisors. One of their tasks was to define the new planning guidelines for the Bijlmermeer on the basis of the working group's report.

The basic problem that triggered the need for a renewal of the Bijlmermeer is summarized in a letter from the Steering Committee:

The issue that prompts this operation is the uncontrollability of the Bijlmermeer, as expressed in its disarray, vandalism and unsafety. This causes a high turnover of residents and a high rate of vacant apartments, resulting in huge exploitation losses. This problem is the immediate result of the district's town planning structure, marked, on the one hand, by homogeneous housing in the form of rental apartment buildings for families of average means, and, on the other hand, by expansive public and semi-public spaces that require an excessively large maintenance crew.[86]

Clearly, the Steering Committee abandoned the limited definition of the Bijlmermeer problems as a housing issue, and defined its particular problems in a much broader light by also taking into account social problems, such as the large number of refugees and migrants in the Bijlmermeer and the high unemployment among its residents.[87] The plans for the neighborhood's renewal meant a distinct break with previously applied standards. Conceptual principles, like the segregation of traffic types and the uniform daily living environment, were not taken for granted anymore. In many respects, the new plans even suggested a radical departure from the tradition of the Bijlmermeer's original planning structure. (See table 4.1.)

The Steering Committee made a distinction between three types of renewal: spatial, social, and managerial. Moreover, the Bijlmermeer was divided into two "focus areas": Ganzenhoef and Amsterdamse Poort.[88] The neighborhood authorities decided that the Bijlmermeer would be renewed step by step, one area after another. Ganzenhoef was chosen as the first focus area because this section's problems were worst. Moreover, some elements of the original Bijlmermeer structure were already redesigned in this area. Therefore, it would be easier to bring about more differentiation, and the neighborhood board expected less resistance from the residents.[89] Demolition of 25 percent of the apartment buildings continued to be the Steering Committee's main goal. For Ganzenhoef this meant the demolition of two buildings. In addition, the committee proposed to lower the Bijlmerdreef, the main road in the Bijlmermeer.

Table 4.1 Comparison of leading themes in the original concept of the Bijlmermeer district and the plans of the 1990s.

Original plans	Plans of 1990s
Much public space, large green areas	Concentration of buildings, less open space and more controllable open space
Car-free public space	Car-centered public space
Egalitarianism	Diversification, differentiation
85% high-rise apartment buildings	Mixture of low-rise, mid-rise, and high-rise buildings
Segregation of traffic types	Mixing of traffic types
Separation of living, working, recreation, and traffic	Mixing of functions

The neighborhood board was divided on the renewal plans. In July 1992 the newly developed planning guidelines for Ganzenhoef were discussed in the neighborhood council. It appeared to be very difficult to win the council's support for this plan. The representatives of Groen Links, a leftist political party in both the City Council and the neighborhood council, argued against the proposals for the demolition of two high-rise apartment buildings. They were afraid that the demolition of two of the buildings would merely displace the problems to the remaining apartment buildings. They commented that the demolition plans were premature because they had not passed the City Council yet, and they also doubted the view that there would be no demand for the apartments. To substantiate their point, they showed statistics that indicated a recent lowering of both the vacancy rate and the turnover rate.[90]

The neighborhood board insisted on the *social* renewal dimension of the plan, but it had great difficulties with the drastic nature of the *spatial* renewal plan. Former neighborhood board member and chairman of the neighborhood council Ronald Janssen, a member of the PvdA party, argued that spatial renewal was seen as a precondition for social renewal. The board was not against demolition, but it wanted to have more scenarios than the one presented in the Ganzenhoef plan. However, according to Janssen, the neighborhood council was pressured by the city to accept the proposed plan.[91] Finally, it indeed was accepted, but, in response, two City Board members (from Groen Links) stepped down. Janssen explained the council's decision with reference to the momentum that was building at that time, a situation from which it could not escape anymore. The start of the process was more important than the specific results in the Ganzenhoef area, so the council felt it needed to accept the plan.[92] Its decision seemed to have important ramifications, though, because it set the stage for further plans and the subsidies that were requested. For instance, the Central Housing Fund supported the renewal with a large subsidy, but this generosity, in turn, made it difficult to divert from the accepted plan.

It might now seem as if in the early 1990s the Bijlmermeer had become entirely malleable, subject to whatever urban redesign schemes actors could think of; after all, demolition had become a real option and the basic assumptions of the newly proposed plans seemed to contradict some of the original ideas behind the Bijlmermeer. (See table 4.1.) Were there also some elements of the Bijlmermeer that resisted change and maintained their obduracy, despite all the radical proposals?

Preserving the Obduracy of the Semi-Elevated Roads

The proposed interventions, particularly the lowering of the roads and the demolition of some of the apartment buildings, were controversial. The critique was often couched in terms of the new plans' deviation from the planning tradition in which the Bijlmermeer was originally conceived. Residents tried to argue the obduracy of the Bijlmermeer structures by invoking aspects of the CIAM tradition, in this case the separation of traffic types. For example, the Werkgroep Wonen en Woonomgeving (Housing and Living Environment Working Group) of the Stichting Wijkopbouw Orgaan Bijlmermeer (Bijlmermeer Community Workers Organization) saw the semi-elevated roads in the Bijlmermeer, including the Bijlmerdreef, as crucial to the original Bijlmermeer design, and therefore they opposed the plans to lower the Bijlmerdreef. According to many, the road infrastructure of the Bijlmermeer was even more fundamental to its design than, for instance, the high-rise apartment buildings. As architect Pi de Bruijn argued: "If you change that [the road system] and preserve the honeycombs, then it doesn't matter what you do, you've still wrecked the Bijlmermeer."[93]

Dirk Frieling, however, considered the lowering of the Bijlmerdreef one of the most crucial redesign interventions. In his view it was more important to do something about the sinister spaces beneath the Bijlmerdreef or the badly functioning service areas than to revitalize the high-rises. Upgrading the service areas to nice, safe, public domains would strengthen the identity of the neighborhood. Strategically, Frieling wanted to take the attention away from the discussion about the demolition of the apartment buildings and shift it to the lowering of the roads. This strategy was not successful, though, for a number of Bijlmermeer residents saw this effort as undermining another important element of the Bijlmermeer concept.[94] An alternative solution for the ground floor's safety problem was to "compress" the area by building low-rise houses in the green areas near the apartment buildings. Moreover, it was proposed to mix the various flows of traffic: cyclists as well as pedestrians were allowed on the elevated roads that were initially reserved for motorists.

The Werkgroep Wonen en Woonomgeving disagreed with the proposal to lower the Bijlmerdreef as well as with the idea of demolishing apartment buildings. Following the principles outlined in a first report on the redesign of the Bijlmerdreef, they pleaded against abolishing the separation of slow and motorized traffic because it did not comply with the "basic design" of the Bijlmermeer: "We want to hold onto both the Bijlmerdreef at its present semi-elevated level and the central area at its present ground-floor level because this characteristic setup determines the way the Bijlmermeer looks. Lowering the Bijlmerdreef would strike at the roots of the basic design."[95] Eventually the working group proposed its own, considerably less expensive plan for the road infrastructure of the Ganzenhoef area in which only a small part of the Bijlmerdreef would be lowered. It pointed out that full reconstruction would cost a lot of money that might well be spent for other purposes. The working group was also concerned about decreased traffic safety after lowering the Bijlmerdreef:

"Demolition of the whole Bijlmerdreef does not provide a solution in terms of livability and safety. On the contrary, the construction of a level road is detrimental to what has been considered as one of the strongest advantages of this neighborhood for years: traffic safety."[96] The plan was not approved; the neighborhood council had already approved a structural plan in which the lowering of the Bijlmerdreef was proposed.[97] However, the working group's views on this issue caused the lowering of other elevated roads in the Bijlmermeer to remain a controversial matter. Traffic engineers, for instance, emphasized the advantage of the elevated roads in terms of traffic safety,[98] but from the angle of social safety the three partners in charge of the renewal (the housing corporation, the neighborhood board, and the City Board of Amsterdam) favored lowered roads.[99] Those who made an effort to preserve the elevated roads did not succeed by mobilizing the planning tradition of the Bijlmermeer. Some elevated roads were nevertheless preserved, but financial arguments and the roads' embeddedness in the overall traffic structure and planning setup (e.g. the accessibility of buildings and indoor car parks from the elevated roads) were considered more important than the argument that the elevated roads were a fundamental part of the Bijlmermeer's basic form.[100]

Demolishing Apartment Buildings

In the discussions about the demolition of a number of apartment buildings in the Bijlmermeer, two assumptions of the utopian tradition played a crucial role: that a better society can be made and that a cohesive community can be established by means of planning interventions. According to the Steering Committee, the renewal's main goal of making Nieuw Amsterdam more financially healthy was to be achieved by creating more stability and social cohesion and by strengthening the identity of the Bijlmermeer, including its public image.[101] This would result in an increase of the occupancy rate as well as of the duration of rental periods, which in turn would lower the deficits of Nieuw Amsterdam.[102] This goal was to be reached by a combination of social and technical measures—demolition being one option.[103]

Figure 4.4
Lowering part of Bijlmerdreef (1996). Source: Archive stadsdeel Zuidoost. © Publi Art, Schoorl.

It was argued that, due to the lack of a differentiation between types of apartments, those Bijlmermeer residents whose financial situation improved could not make a "housing career" in the Bijlmermeer, so they moved to low-rise single-family houses elsewhere in the region around Amsterdam or in Almere or Lelystad. According to the Steering Committee, the cultural identity of the Bijlmermeer could be stabilized and strengthened if residents were to stay on in the neighborhood and not move elsewhere. Therefore, the committee proposed the building of a greater variety of homes, notably more low-rise single-family houses with gardens. It reasoned that ensuring the availability of this type of housing would induce more solidly middle-class residents to stay in the Bijlmermeer and might also attract such people from elsewhere. This, in turn, would enhance the district's stability and social cohesion.[104] The occupancy rate would go up, and hence the rental income would increase. More important, perhaps in the long run, an enhanced community spirit would substantially lower the management costs associated with apathy and vandalism.

Although members of the Werkgroep Wonen en Woonomgeving emphatically shared the value of a strong community spirit, they felt that spatial renewal would cause the Bijlmermeer's multi-ethnic society to disintegrate. The demolition of two high-rise apartment buildings in Ganzenhoef was sharply condemned by the working group: "The demolition of Geinwijk and Gerenstein is in fact the destruction of the unique multi-ethnic society."[105] In a report, the working group suggested that the authorities wanted to replace the present (largely poor and black) population with rich whites. It also argued that the Steering Committee merely wanted to free a substantial piece of land so that expensive new homes could be built on it.[106]

Emphasizing that many residents had begun to enjoy living in the Bijlmermeer, the working group fervently disagreed with the idea of demolition as the basis of the renewal plans: "Many residents love the Bijlmer and they are therefore very emotionally involved in everything that means a threat to its continued existence."[107] More specifically, it acknowledged the great value some residents attach to the Bijlmermeer's original planning concept, in particular the huge green areas and its unique urban spaciousness, as evidenced by the distances between apartment buildings and the great views from the top-floor apartments. The working group blamed the local authorities for not showing much sensitivity to these exceptional features. It strongly favored a threefold approach: the completion of the Bijlmermeer's design as originally planned, pragmatic solutions to the management concerns and the launching of a sustained effort aimed at social renewal. Only if specific problems persisted should demolition be investigated.[108]

The working group's plea failed to garner sufficient backing, and in July 1992 the neighborhood council ratified the proposal to demolish 25 percent of the high-rise apartment buildings. However, in October 1992 the debate was strongly influenced by an unexpected event. A big cargo jet crashed in the Bijlmermeer, killing 41 people (most of them crew members and residents) and partly destroying two apartment buildings. Martin Mulder, at that time director of the Project Bureau for the Renewal of the Bijlmermeer, summarized the reactions to the crash:

> The Bijlmermeer crash has been extremely important. After this, half of the people said "This is so sad, we should stop the demolition process immediately," while the other half said "Now we definitely have to pursue the demolition process." Because of the media attention for the disaster, the whole world suddenly was a witness of the Bijlmermeer problems. But the media also spotlighted the intriguing side of the Bijlmermeer community, the particular value of its multicultural society.[109]

Evidently, the disaster shocked everyone, especially because it proved difficult to ascertain who exactly had lived in the buildings that were damaged. After the disaster, many reports in the press focused on the Bijlmermeer's unemployed residents and its illegal immigrants.[110]

In the wake of the crash, the working group argued in the neighborhood newspaper *De Nieuwe Bijlmer* that the disaster should have concrete implications for the renewal plans. It claimed that the decisions made so far had to be reconsidered, arguing for a period of two years in which the renewal plans would have to be reevaluated: "After the disaster, which, apart from immense sorrow, caused the demolition of nearly 300 apartments, nobody can pretend if nothing has happened. . . . In our view, the decision to demolish Geinwijk and Gerenstein needs to be revoked or postponed."[111] According to Toon Borst, a community worker, the disaster and the resultant attention to the Bijlmermeer in the national press had revealed that members of different ethnic groups succeeded quite well in living together in the Bijlmermeer. The neighborhood should therefore be seen as an example of a multi-cultural society, one that should not be torn apart by rivaling views on the demolition of some of its apartment buildings.[112] In response to the crash, Mulder deferred the discussion for a month, after which he argued in a report to favor a continuation of the planning process.[113]

In this phase, planners, politicians, the housing corporation, and investors gradually came to see demolition as the only solution to the financial and social problems the Bijlmermeer was posing in their opinion. In their view, the Bijlmermeer had failed. They regarded the Bijlmermeer as an unsafe area with many social problems, and they held that radically reconfiguring the spatial structures of the Bijlmermeer was the only way to ensure a better future for its residents and for Nieuw Amsterdam. Therefore, they proposed a number of interventions that implied the abolition of the overall concept of the Bijlmermeer.

Obduracy Explained by the Persistent Traditions of "Community" and "Constructability"

The obduracy and flexibility of the Bijlmermeer in this stage can be understood by referring to the important role of some persistent traditions. The value of striving for a cohesive community in particular was the main tradition that became mobilized by different groups of actors. In the early 1990s, the idea of demolition gained such momentum that it became difficult to stop it for those who opposed it. In the wake of the 1992 plane crash, the appeal to the multi-cultural community spirit in the Bijlmermeer—a spirit that was shared by the residents and the authorities—briefly halted the renewal process and temporarily made the Bijlmermeer more obdurate. Residents tried to make the Bijlmermeer more obdurate after the crash by mobilizing the concept of a cohesive, multi-cultural community. In the arguments of Bijlmermeer residents, the apartment buildings became linked to the multi-cultural identity of the district. In this way, the negative image of the Bijlmermeer, reconfirmed in many press reports after the disaster, was counterbalanced by an image that emphasized the significance of the Bijlmermeer's unique multi-cultural community—a significance that reflected larger cultural changes and anxieties in Dutch society during the early 1990s. Although the proposal to demolish apartment buildings meant a radical break with the planning tradition of the Bijlmermeer, the aim of improving the community's cohesion and identity through concrete interventions is a crucial characteristic of the utopian tradition in planning. The longing for stable community life is still an important underlying element in almost any socio-technical redesign effort. In this way, ironically perhaps, a core value of the utopian tradition persists in the renewal of the Bijlmermeer.

A second core value of the utopian tradition that became mobilized in this stage of the unbuilding process was "constructability": the idea that by means of architectural interventions, a better, more cohesive and ordered society could be created. The idea that the new plans for the Bijlmermeer could strengthen the cultural identity of its residents is an example of this line of reasoning. Many of the other parties involved, such as planners, politicians, and the housing corporation, nevertheless continued to view the Bijlmermeer as unsafe and financially risky. Their approaches suggest an ongoing effort to get rid of elements of the original Bijlmermeer concept by demolishing apartment buildings, lowering elevated roads, adding single-family homes, destroying car parks, and so forth. Thus, for the sake of improving the neighborhood's living conditions, they were constantly looking for ways to challenge the obduracy of its urban design. The Bijlmermeer crash re-opened the renewal debate for a short time. However, a new chapter in the Bijlmermeer's history began when the decision was made to actively pursue renewal again.

INTERVENING IN THE ESTABLISHED PLANNING STRUCTURE (1992–1998)

After 1992, as the renewal effort accelerated and more and more plans were drawn up, the original Bijlmermeer concept continued to be attacked. The structural plan for the renewal of the Ganzenhoef section of the neighborhood became available in 1993. Two years later the demolition of the Geinwijk building started, followed by the Gerenstein building in 1996. Moreover, in 1995 a project group had been set up to work on renewal plans for Kraaiennest, a third area that would be redesigned.

In response to the renewal plans, though, Bijlmermeer residents continued to show their involvement by establishing new working groups or unfolding other initiatives. In 1994, for instance, the Bijlmer Museum Foundation was set up to ensure the preservation of the original concept. In 1996, the Zwart Beraad (Black Council) was founded out of discontent with the way the socio-economic renewal of the Bijlmermeer was organized. In the meantime, planners and politicians who were convinced of the need to rearrange the Bijlmermeer failed to agree on the overall spatial renewal strategy: Should the Bijlmermeer be changed in small steps, one section after the other (ad hoc renewal), or would it be best to work with a new master plan? In this phase, the obduracy of the Bijlmermeer was highlighted in two specific discussions: the debate among planners and politicians on a new master plan and the Bijlmer Museum Foundation's efforts to preserve the original structures.

A New Overall Plan?

The planners, architects, and politicians who became involved in making new spatial plans for the Bijlmermeer could hardly ignore the original planning concept. Planners and politicians agreed that the existing spatial structures had to be changed, but their views of the proper strategies for reaching this goal differed. Some emphasized the need for a new overall plan; others favored dividing the Bijlmermeer into smaller sections that could be changed on a case-by-case basis.

The first concrete renewal operation was the demolition of the Geinwijk and Gerenstein apartment buildings in the Ganzenhoef area, supervised by the planner Donald Lambert. He knew that Nieuw Amsterdam advocated the building of low-rise single-family homes. However, while the planning was underway, Lambert introduced an alternative type of housing: mid-rise apartment blocks. Although it looked like a risky undertaking, Lambert insisted on a smooth transition between the existing high-rise apartment buildings and the new low-rise sections. In his reasoning, motivated by socio-political concerns, the high-rise buildings and the low-rise single-family homes represented two opposite social ranks or ideologies, which, preferably, were mediated by a third type of housing, spatially situated in between the two. (See figure 4.5.) Although Lambert's plan meant a radical break with the Bijlmermeer's overriding tradition of high-rise apartment blocks and egalitarianism, he also tried to establish continuities with the original Bijlmermeer plan by sticking to a large-scale setup, an orthogonal grid structure, and large areas of green public space between the buildings. Lambert regretted, though, that the overall concept of the Bijlmermeer was undermined by the introduction of completely different styles of building, designed by different architects for different parts of the Bijlmermeer. He argued that the city, the neighborhood council, and the housing corporation simply could not reach an agreement on a new structure. These parties had decided not to change the whole area at once and first concentrate on the Ganzenhoef section only, yet this decision was heavily criticized by Lambert, who felt that the "ensemble" of the Bijlmermeer could not be renewed on an ad hoc basis.[114] Thus the value of consistency in planning was strongly advocated by Lambert. When he suggested a plan should be drawn up that could also be applied to a larger part of the Bijlmermeer, "there proved to be no way of sustaining it. Politically there was no inclination to deal with the Bijlmermeer at such a fundamental level. . . . The major fear at that time was monotony, which many felt was the Bijlmermeer's undoing; there was a concern that our new 'design layer' which addressed the Bijlmermeer as a whole, would ring in a new wave of monotony."[115] According to Lambert, it was a political problem. To make the problem manageable, politicians drew a boundary around the Ganzenhoef area. Lambert

Figure 4.5
Low-rise buildings in Gerenstein area (1999). Source: Archive Stedelijke Woningdienst Amsterdam. © Dienst Wonen Amsterdam.

saw this as a very arbitrary move because it cut straight through existing apartment buildings. He characterized it as an insensitive political gesture betraying a lack of concern for what was, in essence, a plan.[116]

From the early 1990s on, some articulated the view that a new overall plan was needed.[117] Yet efforts to develop such a plan seemed far from successful. Many individuals took part in the debate. John Westrik, in an article in the architectural magazine *Archis*, argued that the effort to redesign the Bijlmermeer had "failed to produce a vision, capable of guiding the spatial interventions. . . . The unique concept of the Bijlmermeer deserves a thorough-going urban design strategy which embraces the whole of the Bijlmermeer."[118]

Different reasons for the difficulty of producing a new overall design have been mentioned. According to Dirk Frieling, the radical nature of the Bijlmermeer concept itself was a reason for its endurance: "As for the radical nature of the original Bijlmermeer setup, I am inclined to say that the more one-sided the original concept, the greater its chances of survival. That's logical because the more an idea registers as an all-in scheme, the more likely it is to contain the germs of a regression toward ineffective compromise."[119] Igor Roovers, former project manager of Ganzenhoef also pointed at the difficulty of generating a new concept: "It was very difficult to replace the whole Bijlmermeer by a new concept. . . . The Bijlmermeer was a blueprint, and once it was built, it was hard to escape its concept." Nico Pattiwael, director of the Project Office Renewal Bijlmermeer, argued that it proved to be very difficult to come up with a new overall plan because of the diverging viewpoints of the actors involved. Although they tried to reach agreement on a master plan, it ultimately failed to work. Meanwhile, designated areas were selected in order to be able to proceed with the planning process.[120] Frieling suggested that he did not believe in an overall plan for an area of such size and such a diverse population. It was only possible to provide the rough outlines of the planning scenarios.[121]

Ronald Janssen, at that time chairman of the neighborhood council, confirmed the neighborhood board's fear that the same mistake could be made as in the 1960s, when the Bijlmermeer was built. The board refused, therefore, to apply Lambert's plan for the whole Bijlmermeer. Janssen pointed out that there was a need to be flexible in the Bijlmermeer; obviously, a plan that has only a limited reach can do less damage. Moreover, he favored a division of the Bijlmermeer into smaller districts, each of which would have its own identity. A plan based on this preference (which, paradoxically, might also be understood as a "master plan"—albeit an unpretentious one) would accommodate an endless variety of small-scale planning decisions and flexible design changes.[122] Although City Board member Louis Genet agreed with Lambert's master plan, Janssen succeeded in his efforts to block it.[123]

Another effort to soften the Bijlmermeer's obduracy and to uncouple it from its CIAM moorings was undertaken in 1994. The neighborhood council commissioned the planner and architect Ashok Bhalotra, an associate of Kuiper Compagnons, to design a structural plan for the entire renewal area. In this plan Bhalotra tried to capitalize on the positive image of the Bijlmermeer by defining it as a versatile, colorful, successful community with the potential to become an attractive multi-cultural city. His main objective was to create a common source of inspiration and a sense of "civic pride" for its inhabitants. He conceptualized the Bijlmermeer population as a "border" people—a community living on the boundaries of cities, neighborhoods, landscapes, nationalities, and ethnicities. Bhalotra underscored the significance of distinguishing the Bijlmermeer as a place where multiple cultures, economies, religions, and ambitions exist side by side.[124]

If this image was to flourish, Bhalotra argued, the "old modernist dogmas" would have to be rejected: "The utopian ideals behind the building of the Bijlmermeer belong to the past now: It's time people faced up to that. My theory is that you have to shake the Bijlmermeer to its foundations."[125] Taking the existing structure as a kind of frame for implementing new structural elements, Bhalotra expressed a need for weaving new threads into the Bijlmermeer's urban fabric.[126] Bhalotra underlined the importance of metaphors and language in his plans: "I try to use language and the metaphors in language to free ourselves of dogmatic, technical ways of thinking . . . [and] trivial clichés."[127] He explained that the Bijlmermeer has different meanings for its various inhabitants: "There

are thousands of Bijlmers and thousands of Bijlmermeer narratives. It is interesting to integrate these thousands of narratives and perceptions in such a way that
these thousand Bijlmers still go on existing."[128] But Bhalotra also argued that in
the minds of local authorities and specific groups of residents the Bijlmermeer
concept is psychologically and culturally embedded to such an extent that in the
past it became very difficult for people to accept change. For one thing, a lot of
courage is needed to abandon the icon of modernism, according to Bhalotra.[129]
In an interview in *De Groene Amsterdammer*, he also expressed his fascination for
the modernist tradition in architecture and planning: "Its humanistic ideological
base has always appealed to me."[130]

The renewal debates revealed a deep-seated ambivalence toward "the
Bijlmermeer tradition." Politicians, emphasizing the unsafety of the Bijlmermeer, wanted to break with the holistic planning structure. They had no objections to ad hoc renewal. Donald Lambert favored introducing an entirely new
plan based on differentiation and low-rise houses. Yet Lambert also deliberately
created continuities with the past, for instance, by sticking to an orthogonal
setup, a large scale, and large areas of public space: key elements of the original
Bijlmermeer configuration and the modernist planning tradition in general.
Moreover, he wanted to preserve a degree of totality in the overall Bijlmermeer
setup, a sense of consistency, by applying his concept to the whole Bijlmermeer.
Although Bhalotra's views unambiguously suggest an effort to break away from
the "modernist dogmas," his plan also prioritized achieving a new sense of sol-
idarity and community spirit, in an attempt to move away from the Bijlmermeer's negative stereotype.

Planners, then, had no way of ignoring the original Bijlmermeer concept.
Its daunting presence has been earmarked as one of the reasons for the difficulty
of changing it. The sheer size of the Bijlmermeer largely defined the obduracy of its concept. Some planners and some politicians wanted to replace the
Bijlmermeer's original planning concept with another overpowering concept.
Others engaged in the renewal effort favored a more modest approach: dividing the Bijlmermeer into smaller sections that could be renewed on a case-by-
case basis.

Preserving the Bijlmer Concept: The Bijlmer Museum

The attempts to preserve the obduracy of the Bijlmermeer were further strengthened after the establishment of the Bijlmer Museum Foundation in 1994. This foundation was established by a group of Bijlmermeer residents[131] living in a number of apartment buildings in the G/K area[132] and supported by the resident committees of these buildings.[133] Politicians and planners mostly see this group as a small, white, intellectual, slightly elitist vanguard.[134] Arguing that the planning structure was a subject of debate for the politicians and the planners but not for the residents, the Bijlmer Museum Foundation wanted to preserve the original planning ideas on which the design of the Bijlmermeer was based, as written down in a 1965 report titled Grondslagen voor de Zuidoostelijke Stadsuitbreiding.[135] The representatives of the foundation argued that making changes in the Bijlmermeer's layout would not solve the neighborhood's social problems.[136] Although most considered the Bijlmermeer in terms of its socio-economic issues and safety problems, the foundation singled out a more positive element of the Bijlmermeer, namely the beauty of its design, and defined it as a "museum" that needed to be preserved. Whereas most planners and politicians saw the Bijlmermeer as a completed project that had not lived up to its promise, the Bijlmer Museum Foundation argued that the Bijlmermeer was never really finished and that it should be completed and improved according to the original plan.[137]

The view that social renewal was much more essential than spatial renewal was embraced by the Zwart Beraad. The Zwart Beraad, founded to represent the Bijlmermeer's black residents, had been established in 1996 out of discontent with the way black people were excluded from the renewal effort's decision process. It was bothered by the fact that those involved in the renewal effort—planning, finances, management, and construction—were virtually all white, whereas most of the Bijlmermeer's residents were black.[138] According to Swan Tjoa, a member of the Zwart Beraad and a former secretary of Project Office Renewal Bijlmermeer, the logic behind the renewal plans could be characterized as "spatial determinism," the idea that spatial interventions automatically lead to social solutions. The Zwart Beraad felt that "whoever believes that the socio-economic problem of the Bijlmermeer can be resolved by a few simple interventions, preferably from a spatial point of view, fails to appreciate reality."[139] The Zwart Beraad argued that spatial renewal should be related as closely as possible to socio-economic renewal and better management of the apartment buildings; the socio-economic renewal of the Bijlmermeer should aim primarily at stimulating the emancipatory process of the black people of the Bijlmermeer.[140]

Whereas the Zwart Beraad concentrated on social renewal, the Bijlmer Museum Foundation focused on the preservation of the main characteristics of the original design. In its effort to render the Bijlmermeer more obdurate, the Bijlmer Museum Foundation highlighted specific design features that were derived from the CIAM tradition. Donald Lambert had singled out the large green public spaces and the orthogonal setup as elements of the CIAM tradition that had to be visible in a new master plan, but the Bijlmer Museum Foundation prioritized other aspects. In its view, one of most distinctive features of the Bijlmermeer was the strict separation of slow and motorized traffic; since automobiles were consigned to the semi-elevated roads, the residential areas near the apartment buildings were quiet and safe.[141] The neighborhood's park, designed in the style of English landscape architecture and marked by slopes, footpaths, and watercourses, was also seen as major asset of the Bijlmermeer.[142] Bijlmer Museum proponents stressed that "thanks to the functionalist design, the green areas form a coherent union."[143] The foundation argued against the implementation of traditional residential blocks of streets with cars[144]; it reasoned that cars belonged exclusively on the semi-elevated roads and in the multi-story car parks, which were specifically designed for that purpose.[145] Another important characteristic of the Bijlmermeer concept it wanted to preserve was the string of decentralized service areas, where shops, businesses, parking, and public transportation facilities were concentrated. Furthermore, it favored holding onto the watercourse that runs through the car-free zone and links the neighborhood to two city parks. Finally, the Bijlmer Museum Foundation found it important that, in accordance with the original plans, all residential buildings would continue to be situated within less than half a kilometer from a subway station.[146]

As a result of its efforts, one zone has been identified in which the integrity of the original planning principles of the Bijlmermeer will be fully respected, even if this means restoring existing situations or realizing plans or parts thereof that were never completed. The G/K area is generally seen as the one that is already most consistent with the intentions of the designers of the Bijlmermeer. The Bijlmer Museum Foundation has actively lobbied for creating such a zone and participated in planning the changes that are needed to make it fully compliant with the original intentions.[147] This has led to lengthy negotiations between the Bijlmer Museum Foundation, the housing corporation, the neighborhood council, and some of the other parties about the exact boundaries of the "museum" zone and the proper interpretation of the original planning principles.[148]

In 1994, a decision of the Amsterdam South-East Neighborhood Board included a section on the Bijlmer Museum, claiming that "the demolition of apartment buildings and the building of new houses concerning the area between Ganzenhoef and Kraaiennest would, in principle, not be subject to debate."[149] There was no consensus, however, about either the exact definition of the Bijlmer Museum concept or the exact borders of the area. Problems arose when plans were made for the borders of the area. The neighborhood council decided that near the borders of the zone the design principles could be applied in a "lenient" way.[150] For example, the 1996 plan for Kraaiennest proposed to demolish the Koningshoef apartment building to create space for low-rise buildings, to lower one of the elevated roads, and to demolish car parks. The foundation viewed Koningshoef as a part of the Bijlmer Museum. According to George Munnik, secretary of the Bijlmer Museum Foundation, Koningshoef was, technically speaking, a good building, but it was difficult to maintain, which is why Nieuw Amsterdam wanted it to be demolished.[151] According to the text of an address to the South-East District Council, the demolition of Koningshoef was the result of rental problems: there were many unoccupied apartments, and the turnover was high. It was argued that, in addition, the social safety problem in that area could not be solved if Koningshoef were to be preserved.[152]

The Bijlmer Museum Foundation opposed the demolition of car parks if this implied that cars could access the ground-floor area near the apartment buildings. It stressed once again that the Bijlmermeer's most unique design principle was "the car-free, continuous residential park."[153] It also pointed out that the easy access to public services and public transportation network in the G/K area was a major asset that reflected the "successful" way in which the design principles have been applied in this specific part of the Bijlmermeer.[154] Within the "museum" zone, parking on the ground-floor level near the apartment buildings was not allowed.[155]

Obduracy Explained by the Persistent Tradition of Consistency

How can obduracy be explained in this phase? During most of the 1990s, as we have seen, the tradition of the Bijlmermeer concept played a major role in two discussions: the debate about a new overall master plan for the Bijlmermeer and the discussion about the preservation of the original Bijlmermeer structures in the context of the Bijlmer Museum Foundation. The modernist ideas embedded in the Bijlmermeer structure, such as a car-free living space, separation of traffic types, egalitarianism, large areas of public space, and the importance of public transportation, are still highly valued by a number of Bijlmermeer residents, many of whom are members of the Bijlmer Museum Foundation. As my discussion suggests, the foundation mobilized the Bijlmermeer planning tradition in its effort to prevent changes in the physical structure of the neighborhood that did not comply with the original design principles. If there were problems in the Bijlmermeer, it argued, they were not caused by its design. The foundation expressed its admiration for the beauty and consistency of the Bijlmermeer's original design, offering a splendid alternative to the dominant "bourgeois lifestyle" associated with low-rise single-family homes. As a result of its effort, at least a part of the Bijlmermeer will maintain its obduracy; in this zone adaptations will only be made within the planning tradition of the Bijlmermeer as defined by the original plans.

Not only Bijlmer Museum proponents but also planners, architects, and politicians involved in the renewal process were sensitive to the idea that the consistency of the Bijlmermeer design should be maintained. Lambert and Bhalotra, in their efforts to construct a new plan, clearly struggled with the wish to create new forms of consistency. It is striking indeed that these actors, so critical of the modernist tradition in planning, drew heavily on elements of this tradition when they planned the renewal of the Bijlmermeer. In my view, this phenomenon can be explained by the concept of persistent traditions—focusing on the role of structural and shared values like the utopian tradition in planning. The concept of technological frame would primarily highlight the *differences* in thinking and interacting between local actors and thus has less explanatory power in this case.

OBDURACY AND PERSISTENT TRADITIONS

In the process of unbuilding the Bijlmermeer, the project's modernist planning concept has been defining the parameters of the negotiations to this day. My analysis was geared toward mapping the attempts to either change or preserve the established structures. I will now return to the factor that has turned out to be primarily responsible for constituting the Bijlmermeer's obduracy: the long-term persistence of planning traditions. The adherence to and mobilization of the modernist, utopian tradition in planning influenced the unbuilding process of the Bijlmermeer in specific ways. If chapter 2 addressed how the degree of inclusion of groups of actors in specific technological frames contributed to the obduracy of (parts of) Hoog Catharijne in Utrecht, while chapter 3 showed the role of embeddedness as a major factor in the obduracy of the highway in Maastricht, in this chapter the persistence of traditions turned out to be the most appropriate explanation for the obduracy of suburban socio-technical design in Amsterdam.

Instead of focusing on specific social groups that interact in local contexts, this case has emphasized the role of collectively shared rules and values that transcend local contexts, the role of culturally rooted traditions that derive their strength from the fact that many people share them. This perspective high-lighted the role of culturally embedded, long-term, collective planning traditions that contribute to the obduracy of urban structures in a specific context. I have argued that the specific features of the modernist planning concept and certain characteristics of the utopian tradition that became embedded in this plan, such as constructability,[156] community spirit, and rigorous consistency, kept influencing the discussions, negotiations, and choices in the spatial renewal process of the Bijlmermeer until the late 1990s, long after many of the defining circumstances of the original plan had become obsolete. Although the modernist tradition in planning was already heavily criticized in the 1960s, its principles and dogmas persisted in the design practices and ideals of planners, architects, and residents. I showed how these actors kept "mobilizing" the modernist planning tradition in their arguments to either preserve or demolish specific parts of the Bijlmermeer. In this sense, the developments in the Bijlmermeer can be characterized as "path dependent," since for many the modernist planning tradition remained the dominant trajectory to be followed. It turned out to be difficult to deviate from this path.

First, the persistence of the utopian tradition in the discussions about the renewal of the Bijlmermeer explains why it proved difficult to adapt the Bijlmermeer. The planners and architects involved in the renewal of the Bijlmermeer found it difficult to give up on the idea of an overall planning concept for the Bijlmermeer. They valued the consistency of its rigorous concept. Its radical character has been earmarked as one of the reasons for the difficulty of changing it. Minor adjustments of parts of the Bijlmermeer would still violate the integrity of the concept as a whole.

Second, the utopian tradition in planning helps to explain why architects and planners persist in their preference for master plans. Although architects and planners involved in the renewal of the Bijlmermeer were prepared to abolish the CIAM principles, they tried to create an alternative kind of consistency in their new designs. Lambert deliberately created continuities with some of the guidelines of the original Bijlmermeer design in his plan for Geinwijk and Gerenstein: he opted for an orthogonal large-scale setup and large areas of green public space. The value of consistency can thus explain why certain elements of the original Bijlmermeer plan return in the new plans. Although local politicians were initially against all forms of master planning, they still invited Bhalotra to develop an overall structural plan. Besides the role of consistency and master planning, social ideals, like constructability and community spirit, two important values of the utopian tradition, were also mobilized in the discussions about the spatial renewal of the Bijlmermeer. Social cohesion had to be achieved by more differentiated housing and the neighborhood's cultural identity had to be strengthened by improving the decentralized service areas. Striving for a cohesive community was a goal that was shared by politicians, planners, and residents, even though they differed in their choice of strategies. Residents tried to make the Bijlmermeer more obdurate after the 1992 jet crash by mobilizing the concept of a cohesive, multicultural community. Members of the Zwart Beraad recognized the utopian character of the new plans: they characterized the emphasis on spatial renewal as physical determinism and called attention to the need for social renewal.

Third, some residents view a close connection between the specific aspects of the functionalist tradition in planning and the attractiveness of the Bijlmermeer. This is true, for instance, of the proponents of the Bijlmer Museum. They emphasized that Bijlmermeer residents should appreciate their living environment, its green public spaces in particular. Their concern illustrates the different implications the Bijlmermeer concept may have for various groups of Bijlmermeer residents. Although they cherished the radical nature of the concept as enabling an anti-bourgeois lifestyle, originally, of course, the Bijlmermeer was intended for neat, middle-class families. But from a rhetorical angle their argument was certainly productive, because as a direct result of their effort the original planning concept will be preserved as much as possible in one part of the Bijlmermeer. Furthermore, the utopian tradition played a role as the basis for a rhetoric of preservation of this group of "Bijlmer believers."[157] They strategically mobilized specific characteristics of the CIAM tradition in their attempts to preserve the Bijlmermeer.[158] Their effort has ensured that, at least for the time being, one section of the Bijlmermeer will live on as it was originally conceived.

The central mechanism in the constitution of obduracy discussed in this chapter is the persistence of traditions in planning; in various ways many parties have acknowledged the long-term value of the Bijlmermeer's planning concept, with respect to either particular elements or the design as a whole. The mobilization of persistent traditions can explain why (design) choices are repeated over a long time. The Bijlmermeer case has shown that it is difficult to radically abandon an influential planning tradition like CIAM. This differs from the role of dominant frames in the sense that *all* actors involved—Bijlmermeer residents, planners, architects, politicians—shared the values of the modernist utopian tradition to some extent.

Now that I have analyzed the role of obduracy in three processes of urban socio-technical change, it is possible to discuss the strengths and weaknesses of this approach and to consider the possibility of integrating these three conceptions in one theoretical approach to urban obduracy.

Understanding Obduracy in Urban Socio-Technical Change

My central concern in the preceding chapters was the tension between the effort to change cities and the obduracy of the established urban layout. My aim was to explore and gain more insight into cities as a mixture of fixed, obdurate technologies and flexible, malleable technologies. In my investigation of confrontations between "new" plans and "old" urban structures, I chose to focus on specific urban sites or structures that have become subject to what I referred to as "unbuilding activities": planning and redesign activities that are aimed at changing those elements of cities that for some reason are contested. Assuming that debates about the renewal of urban design are always infused with questions about the extent to which particular urban structures resist change, I argued that the obduracy of these structures would be "tested" in concrete efforts to "unbuild" them.

Three case studies of major Dutch urban redesign projects had to provide answers to questions such as these: How can cities be adapted to accommodate newly conceived ideas and policies? Why do urban structures maintain their obduracy despite efforts at urban innovation? Which strategies do special-interest groups and politicians rely on when seeking to either change what seems solidly in place or hold onto what is contested? To answer these questions, I proposed to use theoretical conceptions of obduracy that were developed in the fields of STS and urban studies. I categorized them as three sets of conceptions of technology's obduracy: obduracy related to actors with a particularly high or low inclusion in frames of thinking and interacting, obduracy explained by embeddedness, and obduracy as constituted by persistent cultural traditions. Each set of concepts focuses on different explanatory mechanisms in the constitution of obduracy. Claiming that obduracy cannot be solely explained by referring to material properties of technologies, I suggested that constructivist approaches—taking the "seamless web" of technology and society as a starting point—do more justice to the interwovenness of the various material and the various social, cultural, political, legal, and economic aspects of the phenomenon of obduracy. This is why I argued for a "symmetrical" approach in which it is not assumed a priori or irrespective of contextual factors whether material or social factors explain the obduracy of specific urban structures. I confronted my empirical investigations with these conceptions of obduracy. I subsequently addressed the role of dominant frames (chapter 2), embeddedness (chapter 3), and persistent traditions (chapter 4). Each case study brought the strengths and weaknesses of these theoretical conceptions of obduracy sharper into focus.

In this final chapter, I will first elaborate on the strong and weak points of each of the three models for studying and explaining obduracy. Next, I will argue that, if we are to arrive at a proper or most refined understanding of the phenomenon of obduracy in urban contexts, these various conceptions of obduracy should be applied in a more integrated way. I will also demonstrate how these conceptions are related to specific "unbuilding strategies" and to strategies intended to preserve obduracy. I will explain how these strategies may be used to inform planning actions and policies in comparable projects outside the Netherlands. Finally, I will reflect on the third concern of this book, which was to draw the attention of STS scholars to the city as a major research site and, vice versa, to acquaint urban scholars with the views and methodologies of STS.

STRENGTHS AND WEAKNESSES OF THE THREE MODELS

In the previous chapters, I demonstrated the applicability of the three conceptions of obduracy by applying them to specific case studies. In this way, I was able to show that these conceptions of obduracy were specific enough to have explanatory power in a single case study. By attempting to apply them to a number of "real-life" cases, it became possible to see to what extent these models needed adjustment, what became visible by using these concepts as heuristics and what remained underexposed. I will now discuss the strengths and weaknesses of the three conceptual models (For a summary of the main strengths and weaknesses, see table 5.1.)

Dominant Frames

In chapter 2, I used the concepts of "technological frame" and "inclusion" to analyze urban change and obduracy. Aibar and Bijker, in their case study of the extension plan of Barcelona, had already shown the usefulness of applying the technological frame concept to urban developments.[1] Architects, planners, civil engineers, and citizens may all be included, in smaller or larger measure, in specific

Table 5.1   Strengths and weaknesses of the three conceptual models.

	Strengths	Weaknesses
Dominant frames	Powerful model for explaining obduracy when two or more relevant social groups are opposing each other and have strong and multiple interactions. The model is symmetrical	Relatively static by focusing on either closed-in or closed-out obduracy. Focus on local level and social groups makes it less attentive to wider, structural and contextual developments.
Embeddedness	Of the three models this concept does most justice to the networked character of cities and the historicity of its structures. The model is the most symmetrical of the three.	Fails to address broader economic structures of power and interests.
Persistent traditions	Long-term structural developments that transcend local groups get more attention in this model.	Material factors get less attention. Urban technology remains a black box.

technological frames. I have specified the elements of the technological frames to include elements that are likely to play a role in urban contexts. Goals, problem definitions, problem-solving strategies, user practices, and tacit knowledge are potential elements of all technological frames. On the basis of the case studies analyzed in this book, a few elements can be added that seem quite specific for cities: the crucial role of historical (planning or architectural) exemplars, planning design methods and criteria (including legal requirements), problems that are associated with the planning structure itself, current spatial policies, financial conditions, and dominant prospective solutions (such as the tunnel in the case of the highway through Maastricht).

My analysis in chapter 2 focused on the interactions and negotiations between various actors—Algemeen Burgerlijk Pensioenfonds, the city of Utrecht, its citizens, the NS railway company, and the Industries Fair—about the plans to redesign Hoog Catharijne. It exposed the emergence of two dominant technological frames that remained stable for a period of roughly a decade. The high inclusion of the city and ABP in their respective technological frames resulted in "closed-in obduracy," while another prominent actor, the city's residents, took Hoog Catharijne as a fact of life or proposed radical alternative designs, such as demolition, outside the two dominant frames (a form of "closed-out obduracy"). A newly appointed planning supervisor, Riek Bakker, succeeded in combining elements of the two technological frames in a new design. As a result of this reconciliation, Hoog Catharijne's obduracy was (temporarily) overcome and the renewal effort could move forward again.

The frames model turned out to be particularly helpful in explaining a technology's obduracy when two (or more) groups are opposing each other and have strong and multiple interactions. The model is sensitive to varieties and differences between social groups, and it distinguishes two types of obduracy (closed-in and closed-out obduracy), depending on the degrees of inclusion of actors in specific technological frames. The model is strong in that it links group interactions to an opening of the black box of socio-technical design. It is thus a symmetric approach that does justice to the social and material aspects of obduracy.

Despite these sensitivities and advantages, I could highlight only a limited number of aspects of obduracy by using this specific theoretical approach. A disadvantage is that the model is relatively static in the sense that obduracy is explained by either high or low inclusion in technological frames. Another disadvantage of the frames approach is that it is always focused on groups, and always emphasizing the local level. This makes it difficult to point at wider "external" or structural factors that play a role in the construction of obduracy. This aspect of the SCOT approach has recently been criticized by Klein and Kleinman: "There is no attention for the ways in which institutionalized social values shape components of technological frames. . . . Bijker never considers the ways in which deeply institutionalized social values shape components of a technological frame or actors' interactions or practices more generally."[2] I agree with Klein and Kleinman that frames "are likely to draw on cultural elements with historical resonances in the society at large," but so far this aspect has never been fully developed in the context of the SCOT model.

By introducing the technological frame concept in the SCOT model, Bijker took one step in the direction of linking technological developments to the wider cultural and socio-political context. But because Bijker chose to focus on widely differing technological artifacts in different historical and geographical contexts, it was hardly possible to develop more explicit connections between technological development and its cultural context. In this book, I chose to limit my focus to urban socio-technical developments in the Netherlands in one specific period of time (1960s–1990s). This made it easier to be specific and precise about the cultural and socio political context of the developments I explored. In addition, in my analysis of obduracy it proved to be rewarding to move beyond Pinch and Bijker's exclusive focus on relevant social groups by considering, specifically in chapter 4, the shared cultural values that are not restricted to a single social group.

Although I have argued that the frames model (despite the disadvantages just mentioned) is best suited to explain obduracy in the unbuilding efforts in Utrecht, alternative explanations for the particular difficulties of changing Hoog Catharijne may be revealed by other theoretical approaches. Close attention for the "embeddedness" of Hoog Catharijne's design brings to light other aspects of its obduracy. Such explanation will stress the interwovenness of heterogeneous elements in urban structures as a major factor in preventing renewal. The embeddedness of Hoog Catharijne can be investigated, for instance, by focusing on its geographic location and its connections to the inner city. The raised walkways that link Hoog Catharijne with the inner city, which have been a significant factor in the obduracy of the complex's original design, may be explained with reference to the way the complex is tied to its surrounding urban infrastructure. Furthermore, its obduracy is also partly explained by its legal embeddedness in a contract between the city and building company Bredero that prohibited physical modifications that would conflict with the company's financial interests associated with the profitability of the mall.[3] Thus, a focus on "embeddedness" would reveal aspects of obduracy that were not explained by the "frames" approach.

The limits of the "frames" approach became more apparent in the case of the highway through Maastricht. In contrast to the Utrecht City Project, technological frames formed a less useful explanation for the existing highway's obduracy. Although it may well be possible to identify different technological frames at work in certain stages of the unbuilding process in Maastricht, these frames were not as dominant and stable as in the discussions about the Utrecht City Project. In the debates on the tunnel option versus the diversion option that took place between 1978 and 1982, the emergence of two technological frames can nevertheless be discerned. Clearly, the actions and reasoning of Rijkswaterstaat were predominantly derived from the "diversion" frame, whereas the city's views and arguments mainly relied on the "tunnel" frame. However, Rijkswaterstaat worked on tunnel designs at this stage as well. The diversion option disappeared from the agenda of both actors after the decision of the Minister of Transportation in 1982. While the "tunnel" frame continued to be important for the city leadership, alternatives were not excluded from the discussions, as I noted in my discussion of both the Infralab procedure and the Design Workplace. This shows that the role of these technological frames in the constitution of the obduracy of Maastricht's highway was not as useful as an explanatory model as in the Hoog Catharijne case.

Embeddedness

The notion of embeddedness as one way to explain obduracy seems particularly appropriate to the city, because it captures a wide range of dissimilar, heterogeneous aspects. For one thing, the interwovenness of heterogeneous elements of various ages and scales is a crucial characteristic of the city. For the purpose of this study I limited the notion of embeddedness to urban examples: embeddedness in planning structures, in a larger traffic system, in spatial policies and in legal norms and regulations.

I have shown that embeddedness has a distinct financial dimension since frequently it is considered too expensive to unbuild urban structures once they have become integrated with other urban elements: changing one element would automatically involve an adaptation of other integrated elements. This is an aspect that is also emphasized in Harvey's work on sunk capital.[4] Transportation networks are particularly difficult to change, because they embody huge capital investments and are, at the same time, very much intertwined with other urban networks.

Nevertheless, this economic embeddedness of the highway in Maastricht is in itself not enough to explain its obduracy. My analysis has shown that user practices can also significantly contribute to increasing the embeddedness of urban structures. Embeddedness may also have an important cultural dimension in the sense that the value attached to buildings of architectural importance often adds to the embeddedness of planning structures.

In chapter 3, the concept of embeddedness turned out to be useful in analyzing the increasing integration of the urban section of the highway in Maastricht's overall traffic system, in the city's larger planning structure and in various policies and legal regulations. The conserving power of embeddedness became most clear in my analysis of how the idea of a tunnel became an integral part of municipal policies, as well as of the ideals, expectations and activities of citizens.[5] I analyzed two different strategies that specifically addressed the highway's embeddedness: (1) deliberately ignoring embeddedness, an Open Plan Procedure technique that prompted a wide range of new design options;

(2) explicitly confronting new design variants with the embedded structures by comparing the pros and cons of the various proposals.

The advantage of embeddedness as an explanation of technology's obduracy is that it fits particularly well to the urban setting where the networked character is obvious. Urban structures become more and more intertwined and connected, and this makes them more difficult to disentangle and change in later phases of a city's development. In this respect it is worth noting that the very fact that the urban structures around Maastricht's highway were longer in place than Hoog Catharijne and the Bijlmermeer may account for some of their embeddedness. This is one of the reasons why the concept of embeddedness is more appropriate as an explanation in this case than in the other two.

The two major conceptions of obduracy of urban socio-technology discussed so far, its ties to "embeddedness" and to "dominant frames," both offer opportunities to the actors involved to disregard obduracy. My first case study (of Hoog Catharijne) suggests that an analysis in terms of technological frames leaves open the possibility that actors can have different degrees of inclusion in technological frames. In my second case study (of Maastricht) I showed that actors may temporarily ignore embeddedness altogether. Moreover, I clarified that the very necessity of having to deal with various forms of embeddedness concurrently in urban redesign projects leads to negotiations about the relative importance of the embedded structures, thus pitting them against each other. Or, put differently, in view of the fact that some embedded structures are more embedded than others, actors always have some space for change, for reconfiguring urban socio-technology.

One of the disadvantages of the embeddedness concept is that it fails to address broader social and economic structures of power and interests. Because of the focus on local actors ("actants") and their efforts to create and sustain durable assemblages of human and non-human actors, the importance of long-term traditions and structural aspects remain underexposed.[6] Actor-network theory in general is criticized "for ceding too much power and autonomy to individual actors, rather than to existing structures of power and interests."[7] In the case of the Maastricht highway, I regularly pointed at contextual developments, including the oil crisis, Limburg's lack of effective lobbying in Den Haag, and the changing political context in the Netherlands. These factors, each of which played an important part in the attempts at unbuilding the highway, were difficult to reconcile with embeddedness as an explanation of the highway's obduracy. Such longer-term contextual and structural developments are better explained by my third model of obduracy: the persistent-traditions model.

The role of persistent traditions may also be explored when trying to account for the highway's obduracy. Such approach would highlight, for instance, the tradition of tunnel engineering and the cultural values and metaphors that are associated with that tradition. In chapter 1, I referred to the nineteenth-century tradition of "putting the less glamorous aspects of civilization underground,"[8] a tradition that has also influenced the effort of twentieth-century architects to solve particular redesign problems. In the discussions about the highway's redesign, the city of Maastricht supported its preference for a tunnel by suggesting it would hide the environmentally damaging flow of cars from view, an option equally favored by residents who advocated the creation of a public park to replace the old highway once it had become obsolete. This "persistent tradition" of trying to hide less attractive urban functions from view in part explains why specific types of design solutions, in this case the tunnel, are proposed and supported time and again. An important episode in the Maastricht case can thus be explained using the persistent-traditions model.

Persistent Traditions

My concept of persistent traditions, understood as a theoretical explanation of obduracy, was inspired by notions such as momentum and cultural archetypes (see chapter 1). I preferred to use a more cultural approach, specifically aimed at the analysis of continuity and change in cities. The cultural aspects of the notion of persistent traditions are visible not only in my focus on longer-term processes, on patterns in design characteristics, on the role of collectively shared rules and values, and on the aggregate level of global context, but also in how all this affects interactions and choices at the local level. The notion of persistent traditions has been made specific to the city by focusing on traditions that are crucial to urban developments, such as the tradition of modernist utopian planning in which constructability, community spirit, and rigorous consistency are crucial.

My analysis in chapter 4 focused on the persistence of the utopian CIAM tradition and on how its mobilization in debates and negotiations about the Bijlmermeer's renewal influenced the obduracy of its urban structures. Because elements of this tradition played a role for all groups involved, it would not have been productive to make a distinction between different technological frames. My analysis revealed how during the various stages of the unbuilding process the actors' repeated recourse to specific design characteristics of the CIAM tradition and shared values of community spirit, constructability, and rigorous consistency affected the obduracy of the Bijlmermeer.

The advantage of the persistent-traditions model in explaining the obduracy of parts of the Bijlmermeer was that it did not focus on single social groups, but on shared values. This allowed me to analyze the influence of certain persistent traditions from which no relevant social group can easily escape. In contrast to the frames model, this model of obduracy highlights shared values instead of differences between groups. This could explain, for instance, how it was possible that the actors in the Bijlmermeer were so critical of the modernist tradition in planning and, at the same time, drew heavily on elements of this tradition when they planned the renewal of the Bijlmermeer. This particular point can be explained by the concept of persistent traditions, whereas technological frames would primarily highlight the differences between actors.

Another advantage of this model is that contextual or "structural" developments get more emphasis than in the other two models. As was mentioned earlier, in each case analysis I pointed at the role of "external" historical developments that are not explicitly brought within any conceptual model but which evidently shaped subsequent planning and decisions. Such contextual developments get different weights in the three models. In the frames model, certain developments (e.g., policy decisions made in Den Haag) may become part of the frames of groups of actors. The embeddedness model pays the least attention to contextual factors; the persistent-traditions model pays the most. When it comes to awareness of contextual and cultural factors, urban scholars may enrich the analysis. Urban historians, for instance, have a tradition of emphasizing cultural factors in the development of cities. Their studies can help to refine case analyses of urban redesign processes.

However, not every aspect of the obduracy of the Bijlmermeer could be explained by reference to persistent traditions. As I remarked in chapter 4, the obduracy of the semi-elevated roads in some parts of the Bijlmermeer can be better understood by looking at their embeddedness in the overall traffic system of the Bijlmermeer and at financial considerations. An analysis of this case in terms of dominant frames would highlight the differences rather than the similarities in the shared values of the groups of actors. Specifically, two technological frames or groups might be identified in the Bijlmermeer case: those who strongly favor the preservation of the Bijlmermeer's original structures (the "Bijlmer believers") and those who view it as an unsafe neighborhood and prefer making radical design changes. In the 1970s, the planners of the Bijlmermeer built up a frame in which a positive view of the Bijlmermeer played a major role, suggesting that the renewal effort had to take place within the parameters of its original design. In the 1990s, one group of residents reinforced this particular frame, but the "unsafe Bijlmer" frame gained prominence at the same time. The degrees of inclusion of various groups in these two frames structured the interactions and negotiations between the actors and explain why some actors either fiercely defended or opposed certain redesign solutions. In contrast to the Hoog Catharijne case, however, opposing technological frames did not result in a deadlock in the planning process. The frames in the Bijlmermeer case were also less stable and were not strictly related to the same groups of actors. This is why the technological frame concept is less productive in explaining the role of obduracy in this case study.

The persistent-traditions model is weakest in the sense that "technology" is not so central. Using this model, the obduracy of the Bijlmermeer could be explained without much focus on the material aspects involved. Thus, technology remains largely a black box in this model. This is in contrast with the other two models where technology plays a more central role and where technological and social (f)actors are treated in a more symmetrical way.

———

Toward a Theory of Obduracy in the Urban Context

My discussion of the strengths and weaknesses of the three models, when applied to a specific real-life unbuilding process, has clarified the intellectual value of the different theoretical frameworks. My analysis has brought forward the problems and advantages of the three approaches. It is now time to look at the possibilities for "synergies" between the three models. As my discussion made clear, in most cases, one model alone cannot sufficiently explain obduracy in the urban context. The three specific conceptions of obduracy discussed in this book each highlight different mechanisms in the constitution of obduracy. But I have shown in this section that, in order to get a thorough grasp of the phenomenon, one will often have to combine two or three models in the analysis. In other words, I have shown that the various conceptions of obduracy that were applied to a single case study (in chapters 2–4) can also be used to explain different aspects of obduracy in the other two case studies. My claim is that the main weaknesses of the models that I discussed above can be overcome by integrating the three models in one analysis of obduracy. Table 5.2 provides a schematic summary of the case studies in terms of the three constructivist conceptions of obduracy.

Having emphasized the differences and the specific explanatory powers of the three conceptions and their applicability to more than one case study, I can now relate the three conceptions to one another more systematically. I will argue that these conceptions of obduracy are compatible enough to enable a more integrated and hence more sophisticated understanding of obduracy in urban contexts: Socio-technical urban structures are always embedded in a broader network of rules, plans, policies, standards, institutions, and norms—a network into which they generally become even more solidly integrated, thus further adding to their obduracy. This process is accompanied by the building of technological frames. Such frames may also come into existence around specific plans in planning processes. The values and meanings attributed to urban socio-technical structures will become part of these technological frames—as is true of preferred design solutions, current policies, and economic considerations—and they will also begin to play a crucial role in the negotiation process between different groups of actors

Table 5.2 Summary of the case studies in terms of the three conceptions of obduracy.

	Frames	Embeddedness	Persistent traditions
Hoog Catharijne/ UCP	1988–1995: ground-level frame (city) versus raised-level frame (ABP) (see table 2.1)	Legal embeddedness of Hoog Catharijne in a contract between Bredero and the city; geographical embeddedness of HC between inner city, industries fair and rail tracks	Historical continuity with the tradition of "brilliant" town planning ideas in Utrecht
Highway through Maastricht	1979–1981: Tunnel frame (city leadership and residents) versus diversion frame (Rijkswaterstaat)	Road's integration in local, national and international traffic schemes; embeddedness in town planning structure and ideas about cultural heritage; embeddedness in legal regulations and design guidelines; embeddedness of non-existent technologies (the tunnel)	Tradition of hiding less attractive aspects of civilization underground (tunnel solution)
Bijlmermeer	Bijlmer believers' frame (since the 1970s, first: town planners and designers of the Bijlmer, later Bijlmer residents such as Bijlmer Museum Foundation; "Unsafe Bijlmer" frame (authorities since 1980s, at its peak in 1990s)	Semi-elevated roads embedded in traffic structure of the Bijlmermeer; Bijlmer structures embedded in zoning scheme (1974)	Utopian CIAM tradition: some of its town planning characteristics and associated norms and values such as constructability, search for community and rigorous consistency

about the obduracy of socio-technical elements. When analyzing the role of obduracy, it is important to consider the degree of inclusion of the various actors in a technological frame. High inclusion will result in closed-in obduracy, whereas low inclusion will lead to closed-out obduracy. In contrast to technological frames, persistent traditions transcend local contexts, and they have a more pervasive cultural influence. Persistent traditions, shared values, and collective norms influence concrete unbuilding processes to the extent that none of the actors can simply ignore what is shared, collective, or persistent.

If the three conceptions of obduracy are compatible and reinforce one another in an integrated analysis, as I just argued, it should be worthwhile to analyze the role of obduracy in processes of urban socio-technical change in single case studies by investigating to what extent the three conceptions provide a useful explanation. One reason why an integration of such diverse models of obduracy is legitimate is that the city is a much larger socio-technical ensemble than the technologies studied earlier in SCOT and ANT research. The sheer size of cities and the wide range of material objects, networks, and structures in them make it possible and justifiable to integrate these models in one analysis. It appears that in analyzing empirical studies the epistemological differences between the different approaches in STS—and even between STS and urban studies—become less acute. Indeed, as I have shown, combining the three heuristics is far more fruitful than artificially keeping up the boundaries between disciplines and perspectives.

Unbuilding Strategies

As this study has elucidated, urban renewal always involves the interplay of stasis and various intertwined, uneven processes of change. Cities are always in flux, always changing, but their embedded urban structures do not automatically change. Although in the past decades automobile traffic significantly increased and environmental norms became stricter, the highway that divides Maastricht in two is still there; apparently, it was able to resist the various redesign efforts that were triggered by the changed local circumstances. Similarly, many people favored a radical redesign of some of the basic urban structures of the

Bijlmermeer after it became a troubled neighborhood, but for various reasons it proved difficult to bring renewal plans into line with the persistent functionalist planning norms and values embedded in its original design.

This study provides more insight into the available strategies for actors who engage in the redesign of complex urban contexts. These strategies are particularly relevant for developing urban policies, and they can help policy makers or citizen groups to determine which strategies may be helpful, or doomed to failure, in a next wave of urban unbuilding. On the basis of my models of obduracy, four "unbuilding" strategies for overcoming the obduracy of urban socio-technology may be identified:

(1)   Actors acknowledge the importance of specific social groups involved in a redesign process and the possibility that their specific, potentially opposite, ways of thinking may result in obduracy; they tend to apply a combined strategy in which redesign and the attempt to influence the rigidity of the dominant frames involved go hand in hand.

(2)   Actors accept that specific urban structures do not exist in isolation and that their obduracy is associated with their embeddedness in a greater network or ensemble; they come up with an approach in which the obduracy of various embedded elements is considered and negotiated.

(3)   Actors recognize the importance of persistent traditions in the constitution of obduracy; they try to reduce the importance of that tradition by proposing a radical break with the past.

(4)   Actors assume that material properties are the most important cause of urban technology's obduracy; they tend to concentrate their redesign efforts on material aspects, but this may lead to physical determinism: the conviction that "social" problems will automatically be resolved by "technical" interventions in embedded urban structures.

Successful unbuilding strategies, it seems, pursue urban renewal at a material level as well as at social, cultural, political, and economic levels. The first strategy mentioned in the list above is marked by a combination of design interventions and decision process interventions that are geared toward both urban redesign and the malleability of the (rigid) viewpoints, values, and stakes of the people involved in the unbuilding process. This strategy works toward achieving consensus, often by compromise. The way Riek Bakker found a solution for the deadlock in the Utrecht City Project is an example of this unbuilding strategy.[9] She simultaneously developed a clear insight into the dominant frames of the actors involved, tried to influence these viewpoints in such a way that the actors became willing to cooperate, and developed a creative new plan that embodied an amalgamation of these viewpoints. Bakker's strategy can be evaluated as a rather successful one, though the malleability that resulted turned out to be only temporary. In cases where obduracy is the result of two (or more) opposing technological frames, having an outsider perform an analysis and formulate a new design can be an effective strategy for change.

The second strategy is exemplified by the procedure used in the reconstruction process of the highway in Maastricht. The Design Workplace had to seek a design compromise that embodied a reconciliation of the viewpoints of the citizens, the city's leadership, Rijkswaterstaat, and the national government without being at odds with legal norms, policies, and zoning schemes. This strategy, though not exclusively geared toward changing urban design, seriously considers the embeddedness of urban structures in urban traffic schemes, user practices, and various policies. The obduracy of all the elements in the overall configuration, including their mutual interrelationship, is negotiated and eventually, albeit provisionally, evaluated on the basis of this strategy. The Maastricht case has shown, however, that such consideration of the redesign effort does not guarantee a successful outcome. Instead of simplifying issues, this unbuilding strategy tends to take the various complexities into account, which in turn may complicate the unbuilding process.

The third strategy, breaking with persistent traditions as a means to overcome obduracy and bring about change, was applied in the case of the Bijlmermeer. Ashok Bhalotra, the man in charge of designing a new structural plan, explicitly claimed that he wanted to reject the dogmas of modernist architecture. Acknowledging that the tradition of modernist architecture was still highly valued by some Bijlmermeer residents as well as by members of the planning and architectural professions, Bhalotra was aware that the best way to overcome the obduracy of the Bijlmermeer's structures was to attack the modernist tradition and to try to diminish its influence. Another way to force a break with traditions is exemplified by a more unconscious strategy applied by those who utilize urban space. Users perform unbuilding activities by making a different use of urban structures than designers anticipated. For example, Bijlmermeer residents started commercial activities in the huge car parks, and homeless people used Hoog Catharijne as a place to sleep. These local practices did question the embedded urban structures and contributed to their malleability.

The last strategy, based on a belief in physical determinism, prioritizes concrete technological interventions, assuming that social change will automatically follow. While few present-day architects, planners, or policy makers would be inclined to embrace the idea of physical determinism, many architectural and planning interventions have implicit or explicit social goals. In my three case studies, livability, social safety, and social cohesion were among the main social goals to be achieved by material interventions. As an unbuilding strategy, this approach was most clearly visible in the Bijlmermeer case, where the urban renewal effort, rendered quite concrete in a new plan, was supposed to contribute in important ways to the resolution of social and economic problems.[10] Apparently, when the main target is social transformation, the utopian adagio of "salvation by bricks alone"[11] easily comes to the fore. The disadvantage of this unbuilding strategy is that it almost exclusively considers the physical side of urban socio-technological structures, neglecting the social and cultural norms and values embedded in them.

STRATEGIES FOR PRESERVING OBDURACY

Apart from unbuilding strategies aimed at overcoming obduracy, it is also important to pay attention to strategies for *preserving* obduracy. What can be done to keep existing urban structures in place? Three closely related strategies can be identified that are often employed to preserve obduracy in cities: emphasizing continuity, age, and investment.

One important strategy for preserving obduracy is to establish continuity with the past by emphasizing the importance of the ideas originally embedded in an urban design. Each of my three case studies provides an example of this.

In the case of Hoog Catharijne, continuity was advanced as an argument for preserving the obduracy of the raised level of Hoog Catharijne and to make other parts of Hoog Catharijne malleable again. Riek Bakker emphasized the "brilliance" of the raised-level design of Hoog Catharijne as a planning concept, thereby trying to persuade the city of Utrecht to embrace the advantages of the established design as well. Bakker capitalized on the value of continuity in the planning history of Utrecht, its tradition of making "brilliant" plans, and in so doing she eventually succeeded in opening up the debate about Hoog Catharijne's innovation.

In the case of the highway through Maastricht, the original planning design of the highway as a boulevard with trees, surrounded by mid-rise, representative buildings visible from the road, was seen as a historical point of reference by the city leadership when a new design for the highway trajectory had to be developed. Also, the deliberate choice in the 1950s to make the highway a part of the ring road around the city center was still advocated in later redesign proposals in which it was argued that local traffic should use the highway.

In the Bijlmermeer, the original functionalist ideas embedded in its design were still valued by a number of residents and by planners engaged in its redesign. They strategically mobilized specific characteristics of its CIAM tradition. Moreover, the Bijlmer Museum Foundation actively tried to establish linkages with the past by claiming the lasting value of the Bijlmermeer's planning principles. Their proposal to transform part of it into a museum proved an effective conservation strategy that raised the obduracy of part of the Bijlmermeer. In all three cases, the original structures were embraced as "historical exemplars" that had to be preserved or rehabilitated. This specific "unbuilding" strategy may serve the interests of citizens action groups, planners, politicians, or other stakeholders in equal measure. This strategy is mainly geared toward preserving the obduracy of embedded urban structures by insisting on their lasting value.

Emphasizing the "age" of a particular urban design is another strategy for preserving obduracy. Stewart Brand once claimed that the older a building becomes the more it is loved.[12] These days, however, material lifecycles are being shortened all the time, and buildings and urban infrastructures tend to become obsolete more and more quickly. In the United States, the material lifecycles of urban objects are generally shorter than in Europe. According to Brand, a building lasts approximately 50 years in Britain, but only 35 years in the United States.[13] The architect Aldo Rossi even went one step further, claiming that "a city changes completely every 50 years."[14] He even suggested that urban transformations are a condition for the persistence of a city: "A city persists through its transformations."[15] This investigation seems to confirm this trend: urban projects that were built in the 1960s already became the object of large-scale redesign efforts in the 1990s. Yet, at the same time, my case studies reveal that relatively young buildings already were tagged as monuments. In the Netherlands, buildings have to be older than 50 years to be able to obtain the official status of monument. However, many younger buildings of special importance are placed on provisional lists of monuments. Most of the apartment buildings along the highway in Maastricht are now about 50 years old, and some of them have already been identified as prospective monuments. Registration as a monument generally implies a high degree of obduracy. The Bijlmer Museum, though not officially registered as a monument yet, fits this strategy as well. Moreover, in nearly all cases, a building or urban structure will be protected from radical modifications by merely calling it monumental in public. The paradox seems obvious here: our accelerated culture transforms urban structures into monuments at an ever shorter interval, but this very phenomenon, in turn, results in preservation rather than change.

A reference to the investments at stake, both in a financial and social sense, may also significantly contribute to the construction of obduracy. According to Brand, "adaptation is easiest in cheap buildings that no one cares about."[16] Economic and financial factors play a very important but ambiguous role in processes of urban socio-technical change. The obduracy that results from vested interests is clearly visible in the Hoog Catharijne case, when ABP, the owner of the mall, was not willing to give up on the raised-level design because of its financial interests. But on the other hand, financial considerations (profit decline) were also an important factor in the decision to start the redesign plans, to make Hoog Catharijne malleable again. In the case of the spatial renewal of the Bijlmermeer, financial reasons (the housing corporation almost went bankrupt) also gave rise to more radical renewal plans. In many cases, however, economic interests are part of larger conflicts of interest.

Investment in social terms may also be a factor in preserving obduracy. When the "mirror trajectory" was proposed in the discussions on the reconstruction of the highway in Maastricht, one reason that this option was eventually abandoned was because this trajectory would cut through a former social problem area. Everyone agreed that the construction of a tunnel there would seriously disrupt this neighborhood's social renewal, a process that had just taken off after much effort of everyone involved, and Rijkswaterstaat therefore decided to leave it untouched. This is an example of a situation in which previous investments aimed at triggering concrete social improvements played a role in maintaining the neighborhood's obduracy.

Implications for Urban Policies and Planning Practices

What can we learn from these strategies in terms of urban policies in the international planning context? This book has made clear that the problem of obduracy in the urban context seems particularly pressing in the Dutch situation, because of the lack of space and because of the extensiveness of Dutch (re-)planning procedures. Of course, one should be careful with simply translating the planning practices in the Netherlands to the situation in other countries. Cross-cultural studies have shown that there are indeed very big divergences between different national planning contexts. The political and legal systems of, for instance, the Netherlands and the United States differ significantly when it comes to planning issues. As I pointed out in chapter 1, local authorities play a much more crucial role in decisions about planning projects in the Netherlands than in the United States. Whereas some degree of central control over planning issues by the national government is accepted in the Netherlands, Americans generally distrust centralized governmental control.[17] And the Netherlands, France, and the United Kingdom differ significantly in regard to instruments for the realization of planning projects. Legal cadres, responsibilities, socio-economic circumstances, and national policies and culture differ significantly between countries. And even within one country there can be huge differences between different locations.[18] This is particularly the case in countries where local authorities have great influence on planning trajectories, as in the Netherlands. But despite this, I think that the three models of obduracy and the strategies discussed above can be applied to planning projects in other countries as well. When applied with enough attention for contextual factors, all three models will be applicable to other unbuilding processes in other countries and cultures.

In general, though, the obduracy problem may be less urgent in the United States than it is in the Netherlands. One of the reasons is that land is much more abundant in the United States. This means that planning activities that meet much resistance in one place can be more easily moved to another. Moreover, in the United States it is not always necessary to replace existing buildings, as is often the case in the Netherlands. Another important reason why the problem of obduracy may be less urgent in the United States than in Holland is related to the dominant public perception of space in both countries: in the United States space is seen as a commodity that is replaceable.[19] As a result of this attitude, there is less attention for conservation and cultural heritage than in Europe, making space less obdurate. This becomes visible once more when we compare the turnover rate of buildings in the United States with that of European countries. In many European countries, where there is a strong land preservation ethic, urban redesign is much more difficult and the issue of obduracy more pressing.[20]

To conclude, the unbuilding of cities does not only involve handling steel and concrete, but it also involves dealing with those ideas, policies, traditions and commitments that have become embedded in the socio-technical layout of cities and that actors keep mobilizing in their attempts to either unbuild or preserve cities. This implies, on the one hand, that planners, citizens, politicians, and action groups who want to change urban space should not underestimate the obduracy of embedded urban structures. But, on the other hand, by developing a clearer insight into the mechanisms that constitute obduracy, actors might improve their strategies for successfully addressing this phenomenon. My analysis of the empirical case studies on the basis of the three heuristics of obduracy and my subsequent discussion of unbuilding strategies and strategies for preserving obduracy, hopefully provide insight into the most effective ways to bring about change or to preserve obduracy, in how to avoid or overcome obduracy.

STS AND THE CITY

In accordance with the general constructivist perspective I have advocated in this study, I have argued that obduracy is a result or outcome of specific, interconnected processes—obduracy being the phenomenon to be explained (explanandum). Obduracy is never an intrinsic property of technology, but technologies are *made* obdurate. This book has shown that obduracy and stability are never permanent but rather ongoing accomplishments. It may take as much effort to keep things the same as it takes to change them. Rather than suggesting that the adoption of a social constructivist perspective implies that there is nothing "real" about technology, MacKenzie and Wajcman emphasize that the idea of the social shaping of technology is "wholly compatible with a thoroughly realist, even a materialist, viewpoint."[21] The benefits of this constructivist approach have been argued for throughout this book. It is important, though, to point at a few differences between existing constructivist approaches of sociotechnical development and my analysis of obduracy in unbuilding processes.

In contrast to earlier STS studies that focused mainly on the process of constructing new technologies, I concentrated on the "re-construction" of existing technologies. As was indicated in chapter 1, this is a crucial difference. In earlier SCOT studies of socio-technical change, the interpretative flexibility of an artifact meant that the artifact was still malleable. Closure resulted in decreasing the interpretative flexibility, after which the artifact became obdurate, if only provisionally. I have shown that discussions, negotiations, and increased interpretative flexibility are ways to put the obduracy of embedded socio-technical ensembles under pressure, but these are not always sufficient to cancel obduracy altogether.[22] I will give two examples. In the 1990s the Bijlmermeer had at least two meanings: it was seen as a functionalist, modernist neighborhood and as an unsafe, crime-ridden, run-down neighborhood. The former was relatively obdurate; after all, the Bijlmer Museum Foundation advocated its preservation, and indeed parts of it remained obdurate. But the other Bijlmermeer, the unsafe neighborhood, was very malleable, a meaning that prompted the demolition of apartment buildings and the Bijlmermeer's transformation into a district with single-family housing. In the case of Hoog Catharijne, two dominant meanings began to figure as elements of two technological frames in the late 1980s. These two meanings remained stable for years, and as a result Hoog Catharijne remained obdurate *despite* its interpretative flexibility. Although different and even contrasting meanings can be attributed to a technology, this docs not automatically imply its malleability. Of crucial importance is whether actors will finally succeed in incorporating the new meaning(s) into a new design that will acquire legitimacy.

Furthermore, on the basis of my study it seems justified to criticize the argument in constructivist technology studies that considers obduracy the outcome of a development that begins in maximum flexibility (see chapter 1). This study has shown that the processes in which obduracy is achieved and contested are in fact much more complex. Obduracy plays a role in various interconnected processes of change. This implies that obduracy is a factor at *all* stages of socio-technical development. My case studies amply demonstrate that the obduracy of elements of established socio-technical ensembles is already negotiated in the planning stage of the redesign effort. When actors finally reach agreement on (for instance) demolishing parts of a city, both obduracy and malleability result. Thus, my study suggests that obduracy is not only the outcome of a construction process but also an integral part of its dynamic.[23] This observation has implications for discussions in STS and urban studies about citizen involvement in urban planning. It has been argued in STS that citizens should be involved in decision making on technological projects in an early stage.[24] This study has shown that citizen involvement in itself does not lead to more or less obduracy, or to more or less effective unbuilding processes. At first sight, the sheer immensity of urban redesign projects that date from the 1960s—of which the Bijlmermeer and Hoog Catharijne are well-known Dutch examples—may suggest the far-reaching malleability of urban structures. Although these plans met with resistance, their initial development and implementation took place with very little direct input from citizens. In the 1970s, however, citizens became more and more involved in decisions about urban renewal. The opposition of citizens, who in many cases try to preserve the obduracy of embedded urban structures, resulted in substantial delays in the decision procedures. The 1990s in particular introduced a host of new models of public participation in the Netherlands.[25] National and local governments tried to involve citizens in decision making about their living environment. On the one hand, these models can be interpreted as ways of dealing with obduracy—as efforts to alleviate it. By involving citizens at an early stage of the decision process, policy makers and politicians hoped to raise their commitment, diminish the opposition to redesign plans, and speed up the mandatory procedures. The recent development and implementation of a variety of public participation models to generate public support for planning interventions in Dutch cities is an example of this consensus orientation in Dutch society. In the Netherlands the political culture also seems more consensus oriented than in, for instance, the United States. On the other hand, planning projects in the Netherlands take decades because of the extensiveness of such procedures. Because obduracy plays a role in all stages of technological development, it does not make sense to focus so much on citizen participation to the early stages, as other constructivists do.

In this book, I introduced the city as a hitherto neglected strategic research site in STS, and I specifically relied on STS concepts to investigate a concrete dimension of urban reality. By emphasizing the heterogeneous character of the city and viewing planning as a mode of socio-technical change, I stressed the reciprocity of the relationship between cities and technologies, conceptualizing them as joint, interconnected categories. The STS notions of "seamless web" and "co-evolution" have been helpful in analyzing the wide array of social, cultural, political, and economic processes of change that concurrently take place in cities all the time. Furthermore, I tried to pry open the black box of the city by focusing on some concrete urban design choices, specific ideas that became embedded in urban layouts, and actual negotiations between various groups of actors engaged in bringing about urban change.

My three case studies illustrate that a focus on various conceptions of obduracy, though originally not intended to be pertinent to *urban* socio-technical change, leads to valuable insights when specifically applied to the city. In this way, this book has shown that STS has a wider applicability than hitherto assumed. STS concepts can also be fruitfully applied to larger artifacts, such as cities. I hope this study will encourage other STS scholars to explore the city as a strategic research site. By the same token, I would be gratified if urban scholars were to be convinced that STS concepts can be especially useful in analyzing processes of urban socio-technical change—a topic architects and urban historians have dealt with from their perspectives. This book has pointed at the possibilities of a productive merger between the two fields of STS and urban studies. STS and urban studies can benefit from each other in the analysis of urban socio-technical change and specifically in research on obduracy. This study meant to explore the possibility of connecting the hitherto largely distinct disciplinary traditions that address urban research and technological developments. In the final analysis, then, this investigation is perhaps most productively understood as a further contribution to the establishment of a common interdisciplinary playing field for these disciplines.

Notes

NOTES

CHAPTER 1

1. The term "unbuilding" was inspired by Donald MacKenzie and Graham Spinardi's notion of uninvention. MacKenzie and Spinardi introduced this notion in their essay on the role of tacit knowledge in the uninvention of nuclear weapons, published in MacKenzie 1995. They argued that the general consensus that nuclear weapons cannot be uninvented rests on a traditional, cumulative view of technological development. Stressing the importance of unrecorded and slowly acquired tacit knowledge in the development of nuclear weapons, they argue that, at least in theory, this knowledge may gradually disappear.

2. Source: "A2-Traverse Maastricht. Geen eindpunt van Nederland, maar startpunt voor Europa zonder hindernissen" (Gemeente Maastricht, 1998) (personal archive of O. de Jong).

3. For examples of the debate in the Dutch media about spatial planning activities in the 1990s, see Hofland 1996; Brusse 1997; Aarden 1997; Huisman 1999. For a publication contributing to the Dutch political debate on spatial planning, see Hajer and Halsema 1997.

4. Huisman 1998. For an overview and discussion of recent urban redesign ("Re-urb") projects in the Netherlands, see Crimson 1997. Examples of Dutch redesign projects in the 1990s are the three projects that are central in this book: the Utrecht City Project, the

reconstruction of the highway that cuts through Maastricht, and the spatial renewal of the Bijlmermeer. Other examples are the reconstruction of the city centers of Enschede, Almelo, Almere, and Den Haag (The Hague).

5. For instance, Das, Leeflang, and Rothuizen (1966) made a plea for the urgent need to develop new, less space-consuming technologies of spatial planning and housing to be able to cope with overpopulation. See also *de Volkskrant* 1997; Blokker 1997; Hendriks and Nijland 1996.

6. Hofland 1996: 5.

7. Konvitz 1985: 188.

8. Other Dutch examples of neighborhoods where CIAM principles were applied are Kanaleneiland in Utrecht, the village of Nagele, and parts of Lelystad.

9. Schuyt and Taverne 2000: 163.

10. Ibid.

11. On the characteristics of the Dutch urban planning system, see Bertolini and Spit 1998, Schuyt and Taverne 2000, and van der Cammen and de Klerk 1993. On the characteristic legal, political, and economic features of the Dutch urban land-use system, see Needham et al. 1993. My analysis of the features American urban planning is mainly based on Perin 1977, Schuman and Sclar 1996, Sies and Silver 1996, and Cullingworth 1997.

12. Perin 1977: 15.

13. Tarr and Konvitz 1987.

14. Schuman and Sclar 1996: 431.

15. On urban reconstruction projects near railway stations, see Bertolini and Spit 1998.

16. Aibar and Bijker (1997) also conceptualize the city as an artifact. Gieryn proposes to view buildings as artifacts, or as "walk-through machines" (2002: 41). In doing this, he aims to make buildings amenable to study with the conceptual tools of STS, originally developed to study technologies.

17. On the "seamless web" metaphor, see Hughes 1988. See also Bijker 1995b.

18. There are a few exceptions, some of which I will discuss below.

19. Johnson-McGrath 1997: 691.

20. Guy, Graham, and Marvin 1997: 193.

21. This distinction between the city as a locus for research and as the focus of research is derived from Hannerz 1980.

22. In *Networks of Power* (1983), Hughes analyzes the introduction of electric power systems in Chicago, Berlin, and London. He shows the importance of the urban context in the shaping of the power systems, emphasizing, for example, that interactions involving technology at the local and regional levels were more important than interactions at the national level. Decisions made by managers at the local level contributed more substantially to the shaping of the systems than the decisions made by engineers or inventors. Between 1890 and World War I, the major electricity companies in Germany, the United Kingdom, and the United States focused on providing electrical power to the largest cities. In Berlin, the city government in its role of regulating agency and the Berliner Elektricitäts-Werke (the private company that supplied most of the city's power until 1915) were the major players in building and shaping the overall power system.

23. See e.g. Mayntz and Hughes 1988; La Porte 1991; Summerton 1994. Summerton 1992 is an exception.

24. Further interesting studies on urban socio-technical developments include Aibar and Bijker 1997, Summerton 1992, Hughes 1983, and Latour 1996. For an explicit attempt to

integrate STS (mainly SCOT) and urban studies in the analysis of urban telecommunication networks, see Graham and Marvin 1996. See also Graham and Marvin 2001.

25. Graham and Marvin 1996, 2001.

26. Gieryn (2002) uses the concepts of interpretative flexibility, heterogeneous design, and black boxing to analyze a biotechnology research lab at Cornell University. Brain (1994) uses ANT to analyze "the way architects responded to the task of translating the social problem of housing into an *architectural* problem in the context of the federally subsidized housing programs of the New Deal." Moore (2001) draws on SCOT, the systems approach, and "critical theory" and applies ANT to the analysis of an architectural project in rural Texas.

27. Burgess 1925; Linder 1996, 1990; Park, Burgess, and McKenzie 1925, 1984. On urban ecological approaches, see Frisbie and Kasarda 1988; Schwab 1992.

28. Frisbie and Kasarda 1988.

29. Park et al. 1925 (1984).

30. Park et al. 1925 (1984): 23.

31. For instance by Gottdiener and Feagin (1988).

32. Burgess 1925 (1984): 52.

33. "Elite groups" will strive for a maximization of the value of land, and as such the land becomes a "growth machine" (Logan and Molotch 1987).

34. Logan and Molotch 1987.

35. Graham and Marvin 2001; Harvey 1985.

36. MacKenzie 1995.

37. Cf. Bijker 1995b.

38. See e.g. Trefil 1994; Lynch 1990 (1958); Vance 1977.

39. In some cases, it may indeed *technically* be very difficult to demolish buildings. In the summer of 1999 Dutch TV showed the fierce attempts to demolish a mausoleum in Bulgaria that was also used as a bombproof shelter. After three failed attempts to demolish the building with dynamite, it was eventually torn down stone by stone.

40. On technological frames, see Bijker 1987, 1995b.

41. For descriptions of the SCOT model, see Pinch and Bijker 1984, 1987; Bijker 1995.

42. Although the concept of closure seems to suggest that technological development is finished and that a technology will exist uncontestedly afterwards, Bijker (1995b) emphasizes that closure need not always be permanent. He introduces the concept of stabilization to stress that interpretative flexibility in fact never stops and that we should consider "degrees of stabilization."

43. Bijker 1995b: 282.

44. Pinch and Bijker 1987: 46.

45. Technological frames are analogous to Thomas Kuhn's paradigms and to Edward Constant's "technological paradigm" (Bijker 1995b).

46. Bijker 1995b: 282.

47. Aibar and Bijker 1997.

48. Ibid.: 6.

49. Ibid.: 17.

---

50. Ibid.: 19.

51. This emphasis on developments along a certain trajectory or path in economic approaches of technology fits better in my third category (persistent traditions).

52. Gorman and Carlson (1990: 156) discuss how Alexander Graham Bell and Thomas Edison invented the telephone. Bell and Edison formulated different conceptions of the telephone. A comparison of the two inventors' mental models reveals differences in their cognitive style: Bell can be seen as a "top-down" inventor who tested several variations and alternatives in his mind, whereas Edison used a "bottom-up" strategy that started from mechanical representations that ultimately led to the telephone. Like technological frames, mental models may be a source of technology's obduracy. Gorman and Carlson suggest that existing mental models can become so constraining that only a relative outsider can move beyond them and develop fresh ideas. Specifically, they refer to the dominant role of the "Reis telephone" in the mental model of the telegraph community—a model that was rejected by the outsider Alexander Graham Bell, who subsequently succeeded in developing an alternative.

53. Ellis 1996.

54. Ibid.: 265.

55. Ibid.: 278.

56. Law 1991: 175.

57. For the notion of "social shaping of technology," see MacKenzie and Wajcman 1999b.

58. Bijker (1995b) distinguishes three increasingly radical interpretations of the seamless web metaphor: (1) Nontechnical factors are important for understanding the development of technology. (2) It is never clear a priori and independent of context whether a problem should be treated as technical or as social. (3) It is not enough to treat the socio-technical as just a combination of social and technical factors; the social and the technical are two sides of the same coin: the technical is socially constructed and the social is technically constructed.

59. The system concept is used in the Large Technical Systems approach, introduced by Hughes (1983). The network metaphor is introduced in the Actor Network Theory, developed by Callon (1986), Latour (1987), and Law (1991). The notion of "socio-technical ensemble" was introduced by Bijker (1995b).

60. On the relationship between his concept of "momentum" and technological determinism, see Hughes 1994.

61. See e.g. Callon 1986, 1987, 1991.

62. Callon 1995.

63. Latour 1988. Latour's analysis is based on a study by Daumas (1977).

64. Latour 1988: 37.

65. Law 1987.

66. Law 1987: 113.

67. Callon 1987.

68. Graham and Marvin 2001: 10.

69. Harvey 1985: 16.

70. Graham and Marvin 2001: 193.

71. Ibid.: 193–194

72. Ibid.

73. Brand 1994.

74. This aspect of the SCOT approach has been criticized by Klein and Kleinman (2002). For a discussion of the criticism (expressed by Hans Radder) that social constructivists disregard the importance of "non-local norms" in the development of technology, see Hamlett 2003. The category of persistent traditions makes clear how norms that "transcend" local contexts play a role in the construction of urban obduracy. This discussion relates to a more general debate in social theory about models that emphasize agency, locality, and contingency, versus other models that emphasize structure, institutions, and persistence. The sociology journal *Theory and Society* is a major intellectual forum for this debate (Battani, Hall, and Powers 1997; Gieryn 2002; Pels 1997).

75. For an overview of more recent and more sophisticated approaches of path dependency, see Garud and Karnøe 2001.

76. So far, this unpublished conference paper is the only available piece of work on the city-building-regime approach. Recently attempts have been made to conceptualize "regimes" in technological development. According to Rip and Kemp (1998), regimes are "a broader, socially embedded version of technological paradigms" and they function as mediators between technologies and the wider technological culture (or, as Rip and Kemp say, "socio-technical landscape"). Rip and Kemp (1990) emphasize the collective character of technological regimes, which makes it difficult for single actors to change the rules that constitute the basis of a regime. Regimes limit the development of alternatives that do not fit into the existing regime.

77. Gullberg and Kaijser 1998: 5.

78. Williams 1990: 206.

79. Kitt Chappell 1989: 372.

80. See the examples discussed in the subsection on dominant frames. Aibar and Bijker emphasize the differences between the groups of architects and engineers, whereas Ellis focuses on the differences between the world views of different professional groups (highway engineers, land-use planners, designers, and so on). Obduracy is constituted by the robustness of these opposite frames.

81. For an overview of similarities and differences between SCOT, LTS, and ANT, see Bijker 1995.

82. Recently SCOT has come to include semiotic traits. See Bijker 1995b.

CHAPTER 2

1. For a historical account of shopping malls in the United States and Canada, see Maitland 1985 (which also includes some European examples).

2. Kooijman 1999: 162.

3. This is illustrated by a number of articles in local and national newspapers and a number of letters from concerned citizens. See e.g. Valk 1973; Anonymous 1973a,b.

4. Anonymous 1973a (my translation).

5. This discussion became particularly vehement in a decision process involving the Vredenburg concert hall. Vredenburg is situated between Hoog Catharijne and the inner city. The architect Herman Hertzberger made a controversial plan for the concert hall that can be seen as an architectural critique of Hoog Catharijne. Instead of functional separation, Hertzberger emphasized multi-functionality. The chaotic low building of the concert hall contrasted sharply with the modernist, functional high-rise buildings of Hoog Catharijne (Buiter 1993).

6. Bredero is the company that built Hoog Catharijne.

7. Valk 1973 (my translation).

8. Haagsma 1976 (my translation).

9. See e.g. Valk 1973; Haagsma 1976; Anonymous 1978.

10. For a historical analysis of the appreciation of concrete as a building material before the 1940s, see Kuipers 1987.

11. Other explanations circulated in those days as well. The difficulty of changing Hoog Catharijne, for instance, was ascribed to the strict terms of the contract between the city and Empeo. As a result, the change in ideas that took place in the 1970s could not be embedded in a transformed plan. It was very difficult for the city to adapt the plan in line with the new insights, if these adaptations did not concur with the profitability of Hoog Catharijne Ltd. The contract implied that any changes in the plan that would not be agreed upon by Hoog Catharijne Ltd. would require such large financial investments of the city, that it would be forced to relinquish this attempt (Bos, Mik, and Versnel 1979). This explanation of the obduracy of Hoog Catharijne is in line with the notion of (legal) embeddedness, which will be elaborated in chapter 3.

12. Buiter 1993.

13. Anonymous 1978.

14. Ibid.

15. Ibid.

16. Maitland 1985.

17. This reconstruction of the process that led to the building of Hoog Catharijne is largely based on an investigation by Blijstra (1969), a study by Boesenkool et al. (1983), a historical overview by Dettingmeijer (1988), a detailed overview of the history of Hoog Catharijne by Buiter (1993), the original Hoog Catharijne plans (N.V. Maatschappij voor Project-ontwikkeling 'Empeo' 1962, 1963), and interviews with A. Feddes (the chairman of the Empeo-team that designed plan Hoog Catharijne) (Bunnik, June 19, 1996) and M. Dendermonde (author of a novel about the history of Hoog Catharijne and of memorial books about Bredero) (Maastricht, December 10, 1997).

18. In this period, population growth was an issue not only in Utrecht but also in many other Dutch cities.

19. Boesenkool 1983.

20. Limiting the number of cars in the city was not considered in those days.

21. For examples of this practice, see Provoost 1996. Provoost's study focuses mainly on the relations between infrastructure development and architecture in the city of Rotterdam.

22. De Jong 1972. After World War II, canals were filled not only in Utrecht but also in many other Dutch cities. In the 1990s, many cities regretted the decisions taken in the 1950s and proposed to restore the canals. On how Maastricht dealt with increasing traffic circulation from the late 1950s on, see chapter 4.

23. Interview with Feddes.

24. Buiter 1993.

25. Interview with Feddes.

26. Buiter 1993.

27. N.V. Maatschappij voor Projectontwikkeling 'Empeo' 1962.

28. Ibid.

29. Blijstra 1969.

30. The name Hoog Catharijne ('hoog' meaning high) refers to the fact that the mall is 5.5 meters above street level.

31. N.V. Maatschappij voor Projectontwikkeling 'Empeo' 1962.

32. Blijstra 1969. Consequently, car traffic would not be disturbed by pedestrians.

33. Dettingmeijer 1988; interview with Feddes.

34. Interview with Feddes.

35. Buiter (1993) described in detail how the Plan Hoog Catharijne was accepted by the City Council of Utrecht on October 10 and 11, 1963.

36. Buiter 1993.

37. Royal Dutch Industries Fair will be abbreviated to Industries Fair in this chapter. In 1995 the Industries Fair's official name was changed to Royal Dutch Jaarbeurs.

38. Buiter 1993; Blijstra 1969.

39. Buiter 1993.

40. Van Esschoten and Kragten 1987b.

41. When I use the word 'partners', I refer to the city of Utrecht, Algemeen Burgerlijk Pensioenfonds (ABP), the Royal Dutch Industries Fair, and NS.

42. Gemeente Bestuur Utrecht, Algemeen Burgerlijk Pensioenfonds, Koninklijke Nederlandse Jaarbeurs, and Nederlandse Spoorwegen. "Tussenrapportage UCP." Utrecht 1989 (Archive Stadsbalie Utrecht).

43. On other station area redevelopment projects in Europe, see Bertolini and Spit 1998.

44. Minutes meeting of the Project Group UCP, September 22, 1988 (my translation) (Archive NS Vastgoed Utrecht).

45. It was important to concentrate pedestrian flows on one level, because if people would use two levels (both the ground floor and the raised level), the presence of people at each level would become marginal. With regard to social safety, the presence of many people was strongly preferred.

46. Buiter 1993.

47. Jacobs 1964 (1961). For a discussion of Jacobs's ideas, see Boomkens 1998: 262–271.

———

48. Jacobs 1964 (1961).

49. The 1972 exhibition at the Van Abbemuseum in Eindhoven about "the street" is one instance of this development. See Deelstra et al. 1972.

50. Minutes meeting of the Project Group, September 22, 1988 (Archive NS Vastgoed Utrecht).

51. Interview with E. Bolt, ABP manager; Heerlen, July 9, 1997. This analysis of "ground-floor thinking" versus an emphasis on the "raised level" can also be more systematically related to specific planning traditions.

52. Gemeentebestuur van Utrecht, Koninklijke Nederlandse Jaarbeurs Algemeen Burgerlijk Pensioenfonds, and NV Nederlandse Spoorwegen. "Het Utrecht-City Projekt. Perspectieven voor de toekomst." Utrecht 1989 (Archive Stadsbalie Utrecht).

53. A letter from the City Board to the city commission for spatial developments and economic affairs (second draft, August 24, 1989) suggests that the reactions to the Perspectives report show that in the discussion, the contrasts between the two models were emphasized inordinately (Archive NS Vastgoed Utrecht).

54. Interview with G. Mik, City Board member UCP (socialist party PvdA) (1990–1994); Utrecht, June 19, 1997.

55. A "revitalization" of the shopping center took place between 1993 and 1995. See below.

56. Interview with Bolt.

57. Interview with D. Regenboog, Rotterdam, September 29, 1997. Regenboog performed different functions during the history of Hoog Catharijne and UCP. He was employed by Bredero from 1969 to 1976. After this, he became involved in the early phase of the UCP as a manager employed by ABP. Then he became external project manager UCP as a Kolpron consultant (1991–1993). Regenboog pointed out that there were also ABP people (including himself) who wanted to combine the ground-floor model and the raised-floor model.

---

58. Interview with H. van Herwaarden, NS Vastgoed; Utrecht, July 1, 1997. Van Herwaarden was also temporarily employed by ABP.

59. On developments in the architecture of Dutch railway stations between 1938 and 1998, see Douma 1998. Utrecht's station hall was enlarged and redesigned in 1989 because of expected capacity problems and in 1996 because of an extension of the number of platforms.

60. "Het Utrecht City Project." Utrecht: N.V. Nederlandse Spoorwegen, May 1989. By C. Douma. Code: 700-446-doc/lm, May 8, 1989/IF (Archive NS Vastgoed). See K. Peters, "Enige opmerkingen bij de nota Het Utrecht City project," March 5, 1989 (Archive NS Vastgoed Utrecht).

61. Interview with P. Nyst, Royal Dutch Jaarbeurs, Utrecht, July 1, 1997.

62. Minutes Steering Committee UCP, April 24, 1991 (Archive NS Vastgoed Utrecht).

63. Bijker (1995b) draws on Hughes's 1983 analysis of the controversy of direct current and alternating current.

64. Bijker 1995b: 279.

65. I described an example of this type of stabilization process in chapter 1. See Aibar and Bijker 1997.

66. In the period 1989–1997, only minor redesign and facelifting activities took place: the redesign of the station hall in 1989 and 1996 and the "revitalization" of the shopping mall's interior design (1993–1995) deserve mention.

67. Gemeente Bestuur Utrecht, Algemeen Burgerlijk Pensioenfonds, Koninklijke Nederlandse Jaarbeurs, and Nederlandse Spoorwegen. "Tussenrapportage UCP." Utrecht 1989 (Archive Stadsbalie Utrecht).

68. Ibid.

69. Ibid.

70. Gemeente Utrecht, NV Nederlandse Spoorwegen, Algemeen Burgerlijk Pensioenfonds, and Koninklijke Nederlandse Jaarbeurs. "Projektplan UCP." Utrecht 1991 (Archive Stadsbalie Utrecht).

71. See e.g. Minutes administrative board UCP, October 12, 1990 (Archive NS Vastgoed Utrecht).

72. Interview with H.S. Yap, planning advisor, UCP (1989–1995); Den Haag, September 28, 1997; telephone interview with A. Hordijk, municipal project manager UCP, initiator of the Atelier, November 18, 1997.

73. Anonymous 1991.

74. Interview with A. Bley, coordinator revitalization, WPM Beheer Midden-Nederland; Utrecht, June 6, 1997.

75. Van Esschoten and Kragten 1987a.

76. Interview with Bolt.

77. Minutes administrative board UCP, December 21, 1992 (Archive NS Vastgoed Utrecht).

78. Gemeente Utrecht, NV Nederlandse Spoorwegen, Algemeen Burgerlijk Pensioenfonds, and Koninklijke Nederlandse Jaarbeurs. "Projektplan UCP." Utrecht 1991 (Archive Stadsbalie Utrecht).

79. Gemeente Utrecht, NV Nederlandse Spoorwegen, Algemeen Burgerlijk Pensioenfonds, and Koninklijke Nederlandse Jaarbeurs. "Projektplan UCP." Utrecht 1991 (Archive Stadsbalie Utrecht). Initially the NS expected an increase from 75.000 passengers a day in 1988 to 130.000 a day in 2010, but the prognoses of March 1990 expected 200.000 passengers a day in 2010.

80. Interview with Bolt.

81. Minutes administrative board UCP, September 12, 1990 (Archive NS Vastgoed Utrecht).

82. Gemeente Utrecht, NV Nederlandse Spoorwegen, Algemeen Burgerlijk Pensioenfonds, and Koninklijke Nederlandse Jaarbeurs. "Projektplan UCP." Utrecht 1991 (Archive Stadsbalie Utrecht).

83. Interview with Mik.

84. Although the UCP became a "potential" key project, it took a long time before it received the official key project status. Leo Lambo, secretary of the BOCP, a citizens' pressure group, argues that the governmental commission that had to decide about the subsidy was not convinced by the plan. Others point out that, because of a change in governmental policies after the elections, key projects came to be considered less effective as policy instruments. However, in May 1999, Deputy Minister Remkes reserved an amount of 100 million to 120 million guilders from the budget for "New Key Projects" for the UCP.

85. Minutes administrative board UCP, March 1, 1993 (Archive NS Vastgoed Utrecht).

86. Stuurgroep UCP. "Master plan Utrecht City Projekt" (Gemeente Utrecht 1993) (Archive stadsbalie Utrecht).

87. Minutes administrative board UCP, March 1, 1993 (Archive NS Vastgoed Utrecht).

88. Interview with Bolt.

89. See "Verklaring behorende bij het Master plan UCP door ABP," April 15, 1993 (Archive NS Vastgoed Utrecht).

90. Mabon/Wilma, MBO, and Multi Development Corporation.

91. Ontwikkelingsmaatschappij Utrecht Centrum Project Bv. "Het ruimtelijk funktioneel concept UCP. Stap I: Evaluatie- en onderzoeksfase." Utrecht 1995 (Archive Stadsbalie

Utrecht). Various organizations and companies collaborated in designing the Spatial-Functional Concept, including project developers, city departments of planning and housing, and private architects' offices.

92. Ontwikkelingsmaatschappij Utrecht Centrum Projekt Bv. "State of affairs" (Utrecht 1994) (Archive Stadsbalie Utrecht).

93. Ontwikkelingsmaatschappij Utrecht Centrum Project Bv. "Het ruimtelijk funktioneel concept UCP. Stap I: Evaluatie- en onderzoeksfase." Utrecht 1995 (Archive Stadsbalie Utrecht).

94. Ontwikkelingsmaatschappij Utrecht Centrum Project Bv. "Het ruimtelijk funktioneel concept UCP. Stap I: Evaluatie- en onderzoeksfase." Utrecht 1995 (Archive Stadsbalie Utrecht).

95. Interview with H. Kernkamp, City Board member UCP (D'66 party) (1994–1998); Utrecht, July 8, 1997.

96. On closed-in and closed-out obduracy, see Bijker 1995a,b.

97. Kiers 1997 (my translation). A. G. Kiers was involved as an engineer in the development of the Plan Hoog Catharijne.

98. This committee united lobbyists and representatives of smaller activist groups from town districts, homeowners, the Union for the Protection of Pedestrians (VBV, Vereniging Bescherming Voetgangers), the Union for Public Transport Passengers (Rover), the League for the Reestablishment of Livability (Herstel Leefbaarheid), the First Dutch Cyclists League (ENFB, Eerste Nederlandse Fietsersbond), and so on. See "Kijk op Utrechts Centrum." Utrecht: Bewoners Overleg City Project 1990 (personal archive of L. Lambo, Utrecht).

99. Ibid.

100. "Standpuntennota BOCP." Utrecht: Bewoners Overleg City Project 1992 (personal archive of L. Lambo, Utrecht).

101. Ibid.

102. Interview with L. Lambo, secretary BOCP; Utrecht, April 29, 1997.

103. "Standpuntennota BOCP," 1992 (personal archive of L. Lambo, Utrecht).

104. Interview with Lambo.

105. Anonymous 1989.

106. Interview with C. Koemans, Utrecht, July 8, 1997.

107. Interview with Koemans. See also "Utrecht Catharijne Park/Stad Projekt. Een visie/plan voor de stad." Utrecht 1993 (Personal archive C. Koemans Utrecht).

108. See "Verwerking externe konsultatierondes (inclusief inspraak) n.a.v. Master plan Utrecht City Projekt." Utrecht: Gemeente Utrecht 1993. chapter 6 (my translation) (Archive Stadsbalie Utrecht).

109. Interview with Kernkamp.

110. Interview with J. Peters, consultant Twijnstra Gudde; Utrecht, June 12, 1997. Interview with A. Lambooy, advisor Riek Bakker; Eindhoven, February 26, 1998.

111. Interview with Lambooy.

112. "Utrecht Centrum Project/Intentie-overeenkomst UCP" 1996/OGU963595 (personal archive of L. Lambo, Utrecht).

113. Bakker 1998: 21–22.

114. Bakker and her company, BVR, were responsible for the urban-planning aspects of the new plan. The consulting company Twijnstra Gudde was in charge of the procedural aspects.

115. Interview with R. Bakker, planning supervisor UCP (1996–1999), Eindhoven, August 14, 1998.

116. Bestuurlijk Platform UCP. "Voorlopig stedenbouwkundig ontwerp." Utrecht: Gemeente Utrecht 1997 (Archive Stadsbalie Utrecht).

117. Interview with Kernkamp.

118. Interview with E. Brandes, planner, city of Utrecht; Utrecht, July 8, 1997.

119. Interview with G. Groener, WBN; Utrecht, June 30, 1997.

120. Interview with Lambooy.

121. Interview with Bakker.

122. Interview with A. Smits, city project manager UCP; Utrecht, June 12, 1997. See also the interviews with van Herwaarden and Kernkamp.

123. Interview with Smits.

124. Interview with Kernkamp.

125. Interview with Kernkamp.

126. Interview with Brandes.

127. This proposal had also been part of the earlier plans (Projektplan, Master Plan, Spatial-Functional Concept).

128. This shift in the definition of Hoog Catharijne fits nicely in the trend described by Dion Kooijman in his study of trends in shopping and the design of shopping malls. He perceives a shift from the "mall as a machine" (the rational and efficient shopping strategy) to the "mall

as a theatre" (i.e., as a recreative space where consumers can shop in a convenient atmosphere) (Kooijman 1999).

129. Interview with Groener.

130. "Herontwikkeling Hoog Catharijne Utrecht." Den Haag: ING Vastgoedontwikkeling B.V. 1997 (personal archive of G. Groener, Utrecht).

131. On mental models, see Gorman and Carlson 1990. On worldviews, see Truffer and Dürrenberger 1997.

132. On the role of concentration and dispersion of groups and the consequences this has for their power position, see Klein and Kleinman 2002.

133. For a discussion of "micropolitics of power" and "semiotic power" in relation to technological development, see Bijker 1995b.

134. However, Yap (2000) doubts whether budgetary problems are the only explanation for the ending of the cooperation between the partners. He suggests that lack of "confidence in the future" is the main cause for the foundering of the UCP in this stage.

135. "Gemeente en NS Vastgoed onderzoeken nieuwe organisatie voor UCP. Samenwerkingsovereenkomst UCP ontbonden" (press release, March 7, 2000); "Voortgang UCP: simpeler, flexibeler en in delen. BenW van Utrecht en NS Vastgoed eens over investeringen in UCP" (press release, May 24, 2000).

136. "Raad stemt in met investeringsprogramma Utrecht Centrum Project" (press release, June 30, 2000) (www.utrecht-ucp.nl).

CHAPTER 3

1. The highway has had several names: (National) Highway 75, European Highway 9 (E-9), (National) Highway 2 (A2, N2), European Highway 25 (E-25).

2. There is an extensive literature on the automobile in American cities. See e.g. Monkkonen 1988; Wachs and Crawford 1991; Lewis 1997.

3. For an extensive and recent review of urban transport literature in both urban history and the history of technology, see Schmucki 2003.

4. McShane 1994.

5. Provoost 1996.

6. On the (social) construction of highways in US cities, see Ellis 1996.

7. Angenot 1948.

8. Different solutions were put forward. For example, an American traffic engineer named Jokinen designed a radical plan for highways through Amsterdam that was never executed (Das et al. 1966).

9. Lange 1948.

10. Rijkswaterstaat has several regional departments, including one in the Province of Limburg.

11. Interview with J. Jamin, Rijkswaterstaat Limburg, highway engineer, involved in the A2 (re-)design process since the late 1950s; Maastricht, February 2, 1999. See also "Notitie voor wethouder openbare werken en sport," December 30, 1974 (Archive Stadsontwikkeling en Grondzaken (SOG), "Rijksweg 75," code 1.811.111/1).

12. Interview with Jamin. See also Minutes City Council Maastricht, June 3, 1958, No. 10-17 (Archive Sociaal Historisch Centrum (SHC) Maastricht).

13. The views of the Chamber of Commerce on Maastricht's traffic problems at that time are illustrated in an advisory report: "Advies uitgebracht door de Kamer van Koophandel en Fabrieken voor Maastricht en Omstreken betreffende noodzakelijke verkeersvoorzieningen

te Maastricht, aan zijne excellentie de Minister van Verkeer en Waterstaat en aan het Gemeentebestuur van Maastricht." Maastricht: Kamer van Koophandel en Fabrieken voor Maastricht en Omstreken 1955 (Archive SHC Maastricht).

14. Interview with Jamin. Interview with P. Jansen, city of Maastricht, traffic engineer; Maastricht, March 4, 1999. Interview with T. Jenniskens, city of Maastricht, expert on Maastricht's history and culture; Maastricht, April 15, 1999.

15. My translation. See Minutes City Council Maastricht, June 3, 1958, No 10-19 (Archive SHC Maastricht).

16. My translation. See Minutes City Council Maastricht, May 7, 1962, No. 6-3 (Archive SHC Maastricht).

17. J. J. J. van de Venne was director of Public Works in Maastricht between 1956 and 1977.

18. Van de Venne 1959: 5.

19. Van de Venne 1958: 2 (my translation).

20. Ibid.

21. Minutes City Council Maastricht, June 3, 1958, No. 10-25 (Archive SHC Maastricht). See also Van de Venne 1964a.

22. Van de Venne 1964b: 18.

23. Minutes City Council meeting Maastricht, June 3, 1958, No. 10-18 (Archive SHC Maastricht).

24. The highway through the city was part of Maastricht's urban expansion plans of 1951 and 1958. The proposal for developing the area around Highway 75 was made and unanimously accepted in the City Council meeting of January 14, 1958. See Minutes City Council Maastricht, January 14, 1958, No. 16 (Archive SHC Maastricht).

25. Minutes City Council meeting Maastricht, June 3, 1958, No. 10-23 (Archive SHC Maastricht).

26. Interview with Jamin.

27. Angenot 1948.

28. Ibid.

29. Berggren 1956; interview with Jamin.

30. Minutes of City Council meeting, Maastricht, June 3, 1958, No.14 (Archive SHC).

31. Thewissen 1958; Bruijnzeels 1960.

32. Van de Venne 1959.

33. Van de Venne 1964a: 165.

34. Essers 1969: 32.

35. Minutes City Council meeting Maastricht, June 3, 1958, No 10-21 (Archive SHC Maastricht).

36. Thewissen 1958.

37. "A2-Traverse Maastricht. Geen eindpunt van Nederland, maar startpunt voor Europa zonder hindernissen." Maastricht: Gemeente Maastricht 1998 (personal archive of O. de Jong, Maastricht). Interview with O. de Jong, city of Maastricht, Department of Town Development, city coordinator of the A2-project; Maastricht, May 7, 1998.

38. Interview with Jamin.

39. Ibid.

40. Bisscheroux and Minis 1997: 80.

41. Archive Stadsontwikkeling en Grondzaken (SOG), "Rijksweg 75," code 1.811.111/1, December 30, 1974.

42. Letter from Rijkswaterstaat to Director of Public Works city of Maastricht, February 1, 1955; see also letter from Rijkswaterstaat to Director of Public Works, March 16, 1955 (Archive Gemeentearchief Maastricht, Archief Afdeling Stadsontwikkeling, nr. 685 "Stukken betreffende aanleg Rijksweg 75 1948–1955," code 1.811.111).

43. Dibbits 1965: 80.

44. In the City of Utrecht, historical canals were filled in the 1960s to create space for traffic (see chapter 2). See Provoost 1996.

45. In Hoog Catharijne and the Bijlmermeer we see examples of this system of vertical segregation of traffic types. See chapters 2 and 4.

46. Bruijnzeels 1960.

47. Rouw 1970.

48. This is not entirely correct when considering the correspondence in 1955 between Rijkswaterstaat and the city I referred to above. In a letter to the city, Rijkswaterstaat argues for a "more spacious" setup of the planning structure surrounding the highway. See letter from Rijkswaterstaat to Director of Public Works of Maastricht, February 1, 1955; see also letter from Rijkswaterstaat to Director of Public Works, March 16, 1955 (Archive Gemeentearchief Maastricht, Archief Afdeling Stadsontwikkeling, nr. 685 "Stukken betreffende aanleg Rijksweg 75 1948–1955," code 1.811.111).

49. "Nota verbetering kruispunt Scharnerweg/Rijksweg 75 (Oranjeplein/Koningsplein)." Maastricht: Rijkswaterstaat directie Limburg, August 1974 (personal archive of J. Jamin Maastricht).

50. Anonymous 1977a,b.

51. See Rijkswaterstaat Limburg, "Nota verbetering kruispunt Scharnerweg/Rijksweg 75 (Oranjeplein/Koningsplein)." Maastricht: Rijkswaterstaat directie Limburg, August 1974 (personal archive of J. Jamin Maastricht).

52. Van de Venne 1964a.

53. Minutes of a Public Council Conference "Beknopt verslag van de op 6 januari 1975 om 20.00 uur gehouden openbare raadsconferentie met als onderwerp Reconstructie kruispunt Scharnerweg/E-9.," January 6, 1975 (Archive SOG Maastricht, "Rijksweg 75," code 1.811.111/1).

54. This observation underlines Harvey's analysis of the fixity of urban structures (especially transportation networks), due to heavy capital investments.

55. "Nota verbetering kruispunt Scharnerweg/Rijksweg 75 (Oranjeplein/Koningsplein)." Maastricht: Rijkswaterstaat directie Limburg, August 1974 (personal archive of J. Jamin Maastricht).

56. "Reconstructie kruispunt E9/Scharnerweg en Geusseltrotonde," August 25, 1975, No. 463 (Archive SHC Maastricht).

57. See chapter 2 for an analysis of how this shift influenced public opinion on the huge urban reconstruction project in the city center of Utrecht: the Plan Hoog Catharijne. The protests against the implementation of the subway in Amsterdam are also notorious in this respect, as well as the fierce actions against the construction of a highway through a nature conservation area near Utrecht: Amelisweerd.

58. "Structuurplan Maastricht 1979." Report no. 168 (personal archive of O. de Jong, Maastricht).

59. For example, the "declaration for the construction of international main traffic routes" (the E-routes, including the E-9), signed in Geneva on September 16, 1950, stated that

E-roads should by-pass built areas when they are led through cities and cause inconvenience and dangers, that E-roads should not have level road junctions, and that intersections and traffic lights should be avoided. These rules, though not compulsory, should be observed as closely as possible. See "Notitie. E-wegen," Rijkswaterstaat directie Limburg, July 13, 1978 (Archive SOG Maastricht, "Rijksweg 75," code 1.811.111/1).

60. "Structuurplan Maastricht 1979." Report no. 168 (personal archive of O. de Jong, Maastricht).

61. Van de Venne 1962.

62. "A2/E9 om en in Maastricht. Tracé studie." Report no. 169. Maastricht: Rijkswaterstaat 1979 (Archive Rijkswaterstaat Limburg Maastricht, afdeling Integraal Verkeer en Vervoer).

63. President Rooseveltlaan, Koningsplein, Oranjeplein, Nassaulaan and Action group B4, Clean Maas Valley, Milieudefensie, Stichting Milieufederatie Limburg.

64. Letter from P. J. G. Groot to the City Council of Maastricht, February 3, 1975 (Archive SOG Maastricht, "Rijksweg 75," code 1.811.111/1).

65. See the appendix to "Central action committee E-9 underground," August 1975 (Archive SOG Maastricht, "Rijksweg 75," code 1.811.111/1).

66. Two reasons were mentioned to explain this: (1) A connection to the Scharnerweg was no longer necessary because the main urban infrastructure connections were already available (namely the Noorderbrug and Europlein); (2) the accessibility of the Scharnerweg was guaranteed via other (projected) local infrastructure connections. See no. 66, January 7, 1981, Archive SOG Maastricht, "Rijksweg 75," code 1.811.111/1.

67. See proposal to the City Council of Maastricht, no. 66, January 7, 1981 (p. 9) and the document that summarizes the opinion of the City Board of Maastricht about the Trajectory Study "Standpunt van de gemeente Maastricht op de hoorzitting gehouden door de Raad van de Waterstaat over het rapport A2/E9 in/om Maastricht," April 29, 1981 (Archive SOG Maastricht, "Rijksweg 75," code 1.811.111/1).

68. Interview with Cremers.

69. See also "Standpunt van de gemeente Maastricht op de hoorzitting gehouden door de Raad van de Waterstaat over het rapport A2/E9 in/om Maastricht," April 29, 1981 (Archive SOG Maastricht, "Rijksweg 75," code 1.811.111/1).

70. Interview with Jansen.

71. See proposal to the City Council of Maastricht, no. 66, January 7, 1981 (Archive SOG Maastricht, "Rijksweg 75," code 1.811.111/1).

72. "Gespreksnotitie t.b.v. periodiek overleg RWS en gemeentebestuur Maastricht op 21 augustus 1979 betreffende studie "E-9 en Maastricht" (Archive SOG Maastricht, "Rijksweg 75," code 1.811.111/1).

73. Interview with A. Lutters, city manager between 1977 and 1998 (Maastricht, July 21, 1999).

74. Interview with J. Kroon, Bouwdienst Rijkswaterstaat, highway engineer involved in Trajectory/EIS study; Apeldoorn, August 18, 1999.

75. "Notitie E-9 en Maastricht ten behoeve van het overleg RWS–Maastricht op 10 april 1978," Rijkswaterstaat Limburg (Archive city SOG Maastricht, "Rijksweg 75," code 1.811.111/1).

76. Letter from Rijkswaterstaat to City Board member Dols, January 27, 1978 (Archive SOG Maastricht, "Rijksweg 75," code 1.811.111/1).

77. Interview with Jamin. I studied Jacques Jamin's personal archive that included numerous highway designs that were made for the A2 by Rijkswaterstaat in the period 1967–1993. Indeed, I found no plans for a tunnel until the "rough" 1981 tunnel designs. It is thus likely that closed tunnel variants were first investigated by Rijkswaterstaat in detail in the Working Group Tunnel Design (established in 1989) and the Trajectory/EIS study (started in 1995).

78. Damoiseaux 1981.

79. "Inspraakreacties inzake de nota A2/E9 om en in Maastricht," November/December 1980 (Archive SOG Maastricht, "Rijksweg 75," code 1.811.111/1).

80. Letter from Council for Water Works to the Minister of Transportation, November 11, 1981 (my translation) (Archive SOG Maastricht, "Rijksweg 75," code 1.811.111/1).

81. "Vaststelling van het tracé van de rijksweg A2/E9 in Maastricht," July 23, 1982 (HW/WWO 39296) (Archive Rijkswaterstaat Limburg Maastricht).

82. Interview with Jansen.

83. Interview with Jansen. On shifts in postwar Dutch governmental policies on traffic and mobility, see Peters 1998.

84. G. van Heusden "Plan van Aanpak Tracé/Mer-procedure Rijksweg 2 Passage Maastricht." Maastricht: Projectbureau MER, February 16, 1995 (Archive Rijkswaterstaat Limburg Maastricht).

85. Interview with Cremers.

86. P. Jansen "Concept. Stand van zaken met betrekking tot de ondertunneling van de A2 te Maastricht." Maastricht: Gemeente Maastricht 1991 (Archive SOG Maastricht, "A2-Traverse Maastricht," code 1.811.111/1).

87. Interview with Cremers.

88. Interview with R. Daniëls, planner, Buro 5, Maastricht, April 21, 1999.

89. Interview with Cremers. Daniëls argues, however, that there was a hidden political agenda behind the City Board's refusal to accept his plan.

90. See e.g. "Maastricht, stad in evenwicht, balans in beweging. Hoofdpunten van het ruimtelijk en economisch beleid 1990–2000," March 1992. (p. 28) and "A2-Traverse.

Voorlopige uitgangspunten en randvoorwaarden Gemeente Maastricht." Confirmed by City Board November 19, 1996 (personal archive of O. de Jong, Maastricht).

91. On the role of legal regulations in planning in the Netherlands, see Van der Cammen and De Klerk 1993.

92. "Schetsontwerp milieubouwplan stadstraverse Rijksweg 75 door Maastricht," November 14, 1978 (personal archive of J. Nakken Utrecht).

93. Anonymous 1983.

94. Interview with Mrs. O. Kars.

95. Bost 1983; interview with Kars.

96. Letter from H. de Groot to City Board of Maastricht, City Council of Maastricht, Rijkswaterstaat Limburg, and the Provincial board of Limburg, July 17, 1987 (Archive SOG Maastricht, "A2-Traverse Maastricht," code 1.811.111/1). Interview with Kars.

97. Tweede Structuurschema Verkeer en Vervoer (Ministerie van Verkeer en Waterstaat 1988).

98. P. Jansen, "Concept. Stand van zaken met betrekking tot de ondertunneling van de A2 te Maastricht," 1991 (Archive SOG Maastricht, "A2-Traverse Maastricht," code 1.811.111/1).

99. Interview with E. Schreuders, Chamber of Commerce Maastricht and region, Maastricht, April 19, 1999.

100. Van der Cammen and De Klerk 1993.

101. Letter from Minister of Transportation to the Dutch Parliament, March 6, 1992 (Archive Rijkswaterstaat Limburg Maastricht, Afdeling Integraal Verkeer en Vervoer).

102. The Trajectory Law concerns the implementation or adaptation of main roads, railroads and water ways. The goal of the Trajectory Law is to combine decision making on infrastructure

with spatial planning procedures and to speed up decision making procedures. For an overview of Dutch spatial planning policies, see Van der Cammen and De Klerk 1993. On zoning schemes and related legal planning instruments, see also Van Zundert 1996.

103. Rijkswaterstaat Directie Limburg. "Tijdelijke maatregelen A2-Traverse Maastricht." Report no. 32. Maastricht: Rijkswaterstaat 1993 (Archive Rijkswaterstaat Limburg Maastricht, Afdeling Integraal Verkeer en Vervoer).

104. Van Zundert 1996.

105. Letter from H. Schroten to M. de Water (Chief engineer-director of Rijkswaterstaat Limburg), July 27, 1992 (Archive SOG Maastricht, "Rijksweg 75," code 1.811.111/1).

106. Minutes of a meeting of Rijkswaterstaat, city of Maastricht, Chamber of Commerce, Province of Limburg, February 4, 1993 (Archive Rijkswaterstaat Limburg Maastricht, Afdeling Integraal Verkeer en Vervoer).

107. The Environmental Impact Statement was introduced in the Netherlands in 1987. Activities that require such a procedure are the development of large-scale neighborhoods, airports, railways, and installations for waste treatment (Cammen and de Klerk 1993).

108. Existing zoning schemes will often be modified as part of a Trajectory/EIS procedure.

109. Callon 1986.

110. "Startnotitie Rijksweg 2-Passage Maastricht. Tracé/mer-studie." Report no. 23. Maastricht: Rijkswaterstaat Directie Limburg 1995 (Archive Rijkswaterstaat Limburg Maastricht, Afdeling Integraal Verkeer en Vervoer).

111. G. Van Heusden "Plan van Aanpak Tracé/mer-procedure Rijksweg 2 Passage Maastricht." Maastricht: Projectbureau MER 1995. (p. 4) Archive Rijkswaterstaat Limburg Maastricht, Afdeling Integraal Verkeer en Vervoer).

112. Letter from the mayor and City Board members of Maastricht to the General Director of Rijkswaterstaat, January 17, 1996, SOG 96-139 (personal archive of O. de Jong, Maastricht).

---

113. Proposal to the City Council of Maastricht, "Voorstel van BenW inzake Startnotitie MER Rijksweg 2-passage Maastricht" February 13, 1996 (Archive SOG Maastricht, "A2 Tracé-mer studie," code 1.811.111/1).

114. The variant above ground level was not investigated in the 1979 study, but there are drawings of Rijkswaterstaat engineer Huisinga of the early 1970s that show this design (personal archive of J. Jamin Maastricht).

115. "Startnotitie Rijksweg 2-Passage Maastricht. Tracé/mer-studie." Report no. 23. Maastricht: Rijkswaterstaat Directie Limburg 1995 (Archive Rijkswaterstaat Limburg Maastricht, Afdeling Integraal Verkeer en Vervoer).

116. Infralab means Infrastructure Laboratory. The first experiments with the Infralab procedure in the Netherlands took place in 1994 and 1995. For an analysis of one of the first experimental Infralab procedures (about the N44 through Wassenaar), and a comparison of the ideas behind the Infralab procedure and the SCOT model, see Hommels 1995, 1997.

117. For a detailed overview of the first two steps in the Infralab procedure in Maastricht, see Pol and Nägele 1996.

118. Interview with J. Magnée, ANWB (General Dutch Automobile Association); Deurne, November 11, 1999.

119. Interview with A. Vesseur, secretary action committee "No east variant for the A2"; Maastricht, April 16, 1999. See Newsletter "Geen Oosttracé voor de A2," September 1996. Another action committee with the same goals was "action committee Amby east." Amby is a city district. In the course of 1997, the diversion was not considered anymore because it would not reduce the amount of traffic at the existing passage. The livability problem at the passage would thus not be solved. In December 1997 it was rumored that the east variant was taken into study again. This was however contradicted by Rijkswaterstaat.

120. Interview with Mrs. O. Kars, resident of one of the apartment buildings near the highway since 1966, member of Interest Group City Passage; Maastricht, July 15, 1999.

121. Interview with Kars.

———

122. Ibid.

123. Ibid.

124. The members of this interest group live in the apartment buildings near the Highway A2: Oranje residence, Oranjestate, President Rooseveltlaan, Nassaulaan, Koningsplein and a small number in city district Nazareth (interview Kars).

125. Interview with Kars.

126. Interview with J. Smeekes, advisor innovation trajectories, inventor of the Infralab procedure; Leidschendam, July 6, 1995.

127. Pol and Nägele 1996.

128. De Rooij 1994.

129. Smeekes, Infrablad special, December 1995. A28 Infrablad, Podium voor co-makers van Infralab. "Hans Smeekes, facilitator: 'Wisselwerking tussen klant en Rijkswaterstaat is enorm zingevend.'"

130. Kune and Van Erkel 1995: 6.

131. See the proceedings of the Infralab procedure in Maastricht, summarized in Pol and Nägele 1996.

132. Pol and Nägele 1996.

133. Interview with F. Hendrikx, Rijkswaterstaat Limburg, project manager A2; Maastricht, February 24, 1999. Interview with C. Konsten, former A2 senior project engineer and chair Design Workplace, Bouwdienst Rijkswaterstaat; Utrecht, April 20, 1999. Interview with Kroon. According to Konsten, the same is true for old trees. In other cases, however, we have seen that old trees lost their obduracy for the implementation of new infrastructure.

134. Interview with Hendrikx.

135. Pol and Nägele 1996.

136. Namely the consulting company Zandvoort Ordening and Advies.

137. 'See Tweede Structuurschema Verkeer en Vervoer (Ministerie van Verkeer en Waterstaat 1988).

138. Minutes of (sub-)Working Group Design, March 30, 1989 (personal archive of J. Jamin Maastricht).

139. See a report by the City Board in which the "provisional starting points and conditions" for the redesign of the city passage are outlined: "A2-Traverse Maastricht. Voorlopige uit-gangspunten en randvoorwaarden." Maastricht: Gemeente Maastricht 1996 (personal archive of O. de Jong, Maastricht). In the minutes of meetings of Rijkswaterstaat engineers, some peo-ple express their concerns about the ambiguity of the municipal guidelines with regard to monuments near the highway. There appear to exist several municipal lists with monuments that have a different status. See Minutes Meeting Bouwdienst/Rijkswaterstaat Directorate Limburg, January 14, 1997, and Minutes meeting Projectteam Bouwdienst, February 4, 1997 (personal archive of J. Nakken, Utrecht).

140. Interview with A. Cremers, Beck, September 14, 1999.

141. Letter from City Board to Rijkswaterstaat, March 12, 1990 (Archive SOG Maastricht, "A2-Traverse Maastricht," code 1.811.111/1).

142. Personal communication O. de Jong, city of Maastricht, Department of Town Development, city coordinator of the A2-project; Maastricht, April 14, 1999.

143. Interview with J. Nakken, Zandvoort Ordening and Advies, chairman Design Workplace; Maastricht, October 28, 1999. Interview with Hendrikx.

144. See an exploratory investigation by the Engineering Department of Rijkswaterstaat: "Rijksweg 2 Passage Maastricht. Een verkenning van integrale ontwerpoplossingen tracé- en verknopingsalternatieven (eerste concept)." Utrecht: Bouwdienst Rijkswaterstaat 1997 (Archive Rijkswaterstaat Limburg Maastricht, Afdeling Integraal Verkeer en Vervoer).

145. See a report by the Engineering Department of Rijkswaterstaat in which the main design variants for the city passage are discussed: "Variantennota." Utrecht: Bouwdienst Rijkswaterstaat 1998 (Archive Rijkswaterstaat Limburg Maastricht, Afdeling Integraal Verkeer en Vervoer).

146. Interview with Nakken.

147. Minutes meeting Project team Bouwdienst, September 22, 1998 (Archive Rijkswaterstaat Bouwdienst Utrecht, codes T-IS, T-AL).

148. Minutes meeting Project team Bouwdienst, September 22, 1998 (Archive Rijkswaterstaat Bouwdienst Utrecht, codes T-IS, T-AL).

149. Minutes Design Workplace No. 21, September 15, 1998 (Archive Rijkswaterstaat Bouwdienst Utrecht, codes T-IS, T-AL).

150. Interview with Konsten.

151. Interview with Hendrikx.

152. Interview with Kroon.

153. See Rijkswaterstaat information bulletin No. 4 and the press release of June 4, 1998 (Archive Rijkswaterstaat Limburg Maastricht, Afdeling Integraal Verkeer en Vervoer).

154. Peters 1998.

155. Interview with Lutters.

156. See Rijkswaterstaat information bulletin No. 5, March 1999 (Archive Rijkswaterstaat Limburg Maastricht, Afdeling Integraal Verkeer en Vervoer).

157. See proposal to the City Council of Maastricht, "Gemeenteraad Maastricht behandelend begroting 1999 en meerjarenraming 2000–2002," December 10, 1998 (personal archive of O. de Jong, Maastricht).

158. See concept decision of the council No. 14, January 5, 1999 Korr. Nr. 98-47416 "Vaststelling Structuurplan Maastricht 2005" (p. 9) (personal archive of O. de Jong, Maastricht).

CHAPTER 4

1. Priemus 1986.

2. Hall 1988: 235.

3. Glendinning 1997.

4. Others say that much research had been done at the time to investigate user needs. See Glendinning 1997.

5. For an analysis of the descrepancies between the plans behind Brasília and how it worked out in practice, see Scott 1998.

6. I will not focus on the decision making process, which has been extensively analyzed by Mentzel (1989).

7. Koolhaas 1977: 3.

8. For an overview of the Modern Movement, see van der Woud 1983. On the role of CIAM in the Netherlands, see Taverne 1983.

9. On shifts in twentieth-century architectural discourse, see Larson 1993.

10. Brain 1994. Brain's analysis of modernist architecture in the United States is particularly interesting because he focuses on the ways architects translate social problems (like housing problems) into architectural problems, and he uses actor-network theory to analyze this.

11. Ghirardo 1996.

---

12. According to van der Woud (1983), Le Corbusier's less nuanced version of the statements of the Athens congress later resulted in much criticism on CIAM as instigator of "rigorous" and "sterile" postwar planning.

13. These two designs of Le Corbusier were never built.

14. Neville-Sington and Sington 1993: 78.

15. van der Woud 1983: 142.

16. Achterhuis 1998. Utopian thinking in has been very influential in Western planning and architecture. Unfortunately, Achterhuis does not pay attention to this in his book. On utopianism in planning and architecture, see Neville-Sington and Sington 1993; Hall 1988; Meyerson 1961; De Geus 1996.

17. It is important to stress here that Le Corbusier's idea of "the city as a machine" is very different from my own position of viewing the city as an artifact, or analyzing the city *as if* it were a machine. My own position is for instance much more 'symmetrical' in the sense that both social and technical elements are taken into account.

18. See also Tod and Wheeler 1979.

19. Neville-Sington and Sington 1993: 76.

20. Ibid.: 37–38.

21. Ghirardo 1996.

22. Scott 1998: 133.

23. This overview of planning principles embedded in the Bijlmermeer structures and the societal and political backgrounds of these ideals and values is mainly based on Bolte and Meijer 1981, Verhagen 1987, Boer and Lambert 1987, Mentzel 1989, Crimson 1997, and the original text of the *Grondslagen voor de Zuidoostelijke stadsuitbreiding* (in which the main norms

and guidelines for the Bijlmermeer plan were laid down): Afdeling Stadsontwikkeling. "Grondslagen voor de Zuidoostelijke stadsuitbreiding. Uitgangspunten en normen voor de stadsuitbreiding Zuidoost." Amsterdam: Gemeente Amsterdam 1965. (personal archive of G. Munnik Amsterdam).

24. Mentzel 1989.

25. See e.g. Dijkhuis et al. 1975.

26. Afdeling Stadsontwikkeling. "Grondslagen voor de Zuidoostelijke stadsuitbreiding. Uitgangspunten en normen voor de stadsuitbreiding Zuidoost." Amsterdam: Gemeente Amsterdam 1965. (personal archive of G. Munnik Amsterdam). This will be referred to as *Grondslagen*.

27. Ibid. (my translation).

28. See *Grondslagen*. The idea of honeycomb ensembles is generally not associated with the CIAM tradition.

29. See Mentzel 1989.

30. See van der Woud 1983.

31. See Das, Leeflang, and Rothuizen 1966.

32. *Grondslagen* (my translation).

33. Verhagen 1987.

34. Bolte and Meijer (1981) mention the following "mistakes" that had been made in the Amsterdam garden cities: too many cars were allowed in the direct living environment of citizens, areas of green public space were too scattered, and services were not concentrated.

35. Bolte and Meijer 1981.

36. Mentzel 1989.

37. Ibid.

38. Verhagen 1987.

39. Neville-Sington and Sington 1993: 79.

40. Hulsman 1999.

41. Neville-Sington and Sington 1993: 79.

42. Bolte and Meijer 1981; Mentzel 1989. Bolte and Meijer conclude that, because of the initial postive response to the Bijlmermeer concept, the plan represented consensus about the model of "the City of Tomorrow."

43. Ter Horst et al. 1991; Mentzel 1989.

44. Larson (1993) identifies a second major shift in architectural discourse in this period: a postmodern reaction against modernist architecture that occurred between 1966 and 1985.

45. See Stoppelenburg 1972. Bolte and Meijer 1981 argue that the planners of the Bijlmermeer have nevertheless tried to incorporate a number of criticisms of functionalism in the Bijlmermeer design. For instance, the ideals on which the "wijkgedachte" (neighborhood unit) was based, return in the emphasis on the house blocks and shared services and spaces. In both cases, the spatial unit is also seen as a social unit.

46. For an analysis of the Bijlmermeer from a planning perspective, see Ter Horst et al. 1991: 4.

47. See Luijten 1997. Others pointed out that the Bijlmermeer was not retrieved by societal development, but on the contrary, far ahead of its time. Interview with D. Frieling, chairman of the Renewal Bijlmermeer Steering Committee (1990–1992); assistant director of the Amsterdam Public Housing Department (1967–1972); Delft, August 18, 1998. Note that

---

many of the interviewees lived for shorter and longer periods in the Bijlmermeer: John Brewster, Pi De Bruijn, Dirk Frieling, Lucas van Herwaarden, Susan van der Hilst, Ronald Janssen, George Munnik, Swan Tjoa.

48. Melger et al. 1987: XII (my translation).

49. Frieling, quoted by Kloos (1997: 23).

50. De Wit 1993: 19 (my translation).

51. Stichting Wijkopbouworgaan Bijlmermeer 1980: 5 (my translation).

52. De Wit 1993: 20 (my translation).

53. Interview with L. van Herwaarden, landscape architect, neighborhood South-East; Amsterdam, July 22, 1999.

54. Interview with P. de Bruijn, architect, employee of the municipal Housing Department (1970–1977), chairman of the Bijlmermeer Management Group (1974 1977); Amsterdam, August 26, 1998.

55. Interview with de Bruijn.

56. See ter Horst et al. 1991. In the Netherlands, planning structures are legally defined in bestemmingsplannen (zoning plans). See chapter 3 on the legal embeddedness of urban structures.

57. Interview with de Bruijn. However, the apartment buildings that were built in the last phase, deviated from the overall planning concept of the Bijlmermeer (ter Horst et al. 1991).

58. See report Stichting Wijkopbouworgaan Bijlmermeer. "Van de Bijlmermeer méér maken. Een deltaplan voor de Bijlmermeer." Report no. 157. Amsterdam: SWOB 1980 (personal archive of P. ten Have Amsterdam).

59. Interview with R. Grotendorst, former employee of the Amsterdam Federation of housing corporations (1980–1984), former (assistant) director of housing corporation Nieuw Amsterdam (1984–1989 and 1989–1996); Amsterdam, August 10, 1998.

60. Melger et al. (1987) criticized this limited definition.

61. See Municipal Housing Department: Gemeentelijke Dienst Volkshuisvesting. "Concept notitie Bijlmermeer." Amsterdam 1970 (Archive stadsdeel Zuidoost Amsterdam).

62. Huls et al. 1983.

63. See Brakenhoff et al. 1991. Grotendorst pointed out that the immigrants were often seen as the cause of the problems in the Bijlmermeer.

64. The cultural anthropologist and sociologist Livio Sansone (1992) studied the activities of Surinam-Creole youth subcultures in the Bijlmermeer. In this culture, "hosselen" is an important activity for the unemployed. To hossel is to undertake activities in the informal economy. This includes a wide range of activities, from earning money by organising small-scale parties and festivals in Bijlmermeer flats and performing odd jobs for neighbors and family to outright illegal activities such as drug dealing and stealing.

65. Interview with Grotendorst.

66. The last measure, admitting cars near the residential buildings, was already in contrast to the ideas of a car free ground floor on which the Bijlmermeer's design was based.

67. See "A Plan for the Bijlmermeer" by the Project Group on High-Rise Buildings in the Bijlmermeer: Projektburo Hoogbouw Bijlmermeer. "Een plan voor de Bijlmermeer." Amsterdam: Gemeente Amsterdam 1983 (Archive stadsdeel Zuidoost Amsterdam).

68. See Renewal Bijlmermeer Steering Committee "Kiezen en beginnen! Eerste werkprogramma van de Stuurgroep Vernieuwing Bijlmermeer." Amsterdam: Stuurgroep Vernieuwing Bijlmermeer 1991 (personal archive of T. van den Klinkenberg Amsterdam).

---

69. Grotendorst (interview) mentioned another reason: a number of planners of the Bijlmermeer were still employed by the city of Amsterdam and would not accept demolition.

70. Interview with Grotendorst.

71. Ter Horst et al. 1991: 33 (my translation).

72. Interview with Grotendorst.

73. Melger et al. 1987.

74. Ibid.

75. See proposal to the City Council of Amterdam, October 26, 1990. No. 891. "De toekomst van de Bijlmermeer." (personal archive of T. van den Klinkenberg Amsterdam and archive Projectbureau Vernieuwing Bijlmermeer). See also Renewal Bijlmermeer Steering Committee. "Werk met werk maken." Amsterdam: Stuurgroep Vernieuwing Bijlmermeer 1992 (personal archive of T. van den Klinkenberg Amsterdam). Interview with Grotendorst; Interview with Frieling. The following parties were involved in this working group: City of Amsterdam, housing corporation Nieuw Amsterdam, District South-East, Ministry of Public Health, Spatial Ordering and Environment, Amsterdam Federation of housing corporations, investors, National Housing Council, Netherlands Christian Institute for Housing.

76. Werkgroep Toekomst Bijlmermeer. "De Bijlmermeer blijft, veranderen." Amsterdam 1990 (Archive stadsdeel Zuidoost Amsterdam).

77. For example, a flat management plan, differentiation of rents, and exchanging flats with other housing corporations.

78. See Werkgroep Toekomst Bijlmermeer. "De Bijlmermeer blijft, veranderen." Amsterdam 1990. (p. 40) (my translation) (Archive stadsdeel Zuidoost Amsterdam).

79. See Office for Metropolitan Architecture. "Revisie Bijlmermeer." Rotterdam: OMA 1986 (Obtained via OMA). This plan has never been executed.

80. Interview with T. van den Klinkenberg, member of the Renewal Bijlmermeer Steering Committee (1990–1992); Amsterdam, May 25, 1998. Interview with R. Janssen, former chairman of the neighborhood council, board member neighborhood South-East (PvdA); Amsterdam, July 31, 1998. Interview with M. Mulder, Director of Project Office Renewal Bijlmermeer (1992–1996); Almere, August 10, 1998.

81. Interview with van den Klinkenberg.

82. See Werkgroep Toekomst Bijlmermeer. "De Bijlmermeer blijft, veranderen." Amsterdam 1990. (p. 43) (my translation) (Archive stadsdeel Zuidoost Amsterdam).

83. Interview with I. Roovers, project manager Ganzenhoef (1993–1997); Almere, May 25, 1998. Interview with Grotendorst. Interview with D. Lambert, planning supervisor for Ganzenhoef, Rotterdam, July 27, 1998.

84. In 1987, the South-East neighborhood was established as a result of the aim to decentralize municipal governance in Amsterdam. South-East comprises the Bijlmermeer and has its own neighborhood council and neighborhood board.

85. Hereafter referred to as the Steering Committee.

86. Letter from Renewal Bijlmermeer Steering Committee to the City Board of Amsterdam, the board of South-East and the Board and members of Housing Corporation Nieuw Amsterdam, May 25, 1992 (my translation) (personal archive of T. van den Klinkenberg Amsterdam).

87. Interviews with Frieling and Grotendorst.

88. Later two focus areas were added: Kraaiennest and Centrumgebied Zuidoost.

89. Notition neighborhood board W+W, Nieuw Amsterdam, April 2, 1991. "Voorzet voor de keuze van een deelgebied" (Archive Projectbureau Vernieuwing Bijlmermeer Amsterdam).

90. See "Stuurgroep vernieuwing Bijlmermeer brengt eindrapport uit" (press report, June 25, 1992) (personal archive T. van den Klinkenberg, Amsterdam).

———

91. Interviews with Janssen and Grotendorst.

92. Interview with Janssen.

93. Cited on p. 27 of van Giersbergen 1997b.

94. Interview with Frieling.

95. Werkgroep Wonen en Woonomgeving Bijlmermeer. "Ganzenhoef-west: Geen vernieling, maar vernieuwing." Report no. 30. Amsterdam: Wijkopbouworgaan Bijlmermeer 1992. (p. 20) (my translation) (Archive stadsdeel Zuidoost Amsterdam).

96. Letter from Working group Housing and Living Environment (Mart van de Wiel) to Commission Renewal Bijlmermeer district South-East and Commission Housing city of Amsterdam. September 19, 1993 (my translation) (personal archive of G. Munnik Amsterdam).

97. Notition Operational staff renewal Bijlmermeer, September 23, 1993 (Archive Projectbureau Vernieuwing Bijlmermeer Amsterdam).

98. Interview with J. Brewster, traffic engineer, neighborhood South-East; Amsterdam, September 22, 1999.

99. Interview with S. van der Hilst, project manager Ganzenhoef; Amsterdam, July 14, 1999.

100. Interviews with Brewster and van Herwaarden.

101. Interviews with Frieling and Mulder.

102. Interview with Frieling. See also "Vernieuwing Bijlmermeer, een vitale operatie." Amsterdam: Stuurgroep Vernieuwing Bijlmermeer 1991 (personal archive of T. van den Klinkenberg, Amsterdam).

103. Although the problems had been defined more broadly, the solutions were eventually limited mainly to technical interventions. According to Martin Mulder (interview), the social

component of the renewal process received much less attention than the physical/spatial component. However, Frieling, in his interview, emphasized that, although the Steering Committee had put much effort in the social renewal, they were obstructed by the local institutions and authorities. According to Frieling, the early efforts of the Steering Committee were more directed at providing jobs for the Bijlmermeer residents than at spatial interventions.

104. Interview with N. Pattiwael, director of the Project Office Renewal of the Bijlmermeer (1997–1998); Amsterdam, June 4, 1998. Interview with A. Bhalotra, Kuiper Compagnons, supervisor of the structure plan Bijlmermeer; Rotterdam, August 18, 1998. Interview with Lambert. On the one hand, the Bijlmermeer population is characterised as "homogeneous, poor and black." On the other hand, its diversity is always emphasized when politicians use the metaphor of the multi-cultural society to point at the positive aspects of the Bijlmermeer.

105. Werkgroep Wonen en Woonomgeving Bijlmermeer. "Ganzenhoef-west: Geen vernieling, maar vernieuwing." Report no. 30. Amsterdam: Wijkopbouworgaan Bijlmermeer 1992. (p. 6) (my translation) (Archive stadsdeel Zuidoost Amsterdam).

106. Ibid.

107. Werkgroep Wonen and Woonomgeving. "Voorwaarts en niet vergeten. Over de toekomst van de Bijlmermeer." Amsterdam: SWOB 1991. (p. 14) (my translation) (personal archive of T. van den Klinkenberg Amsterdam).

108. See Working Group Housing and Living Environment: Werkgroep Wonen and Woonomgeving. "Voorwaarts en niet vergeten. Over de toekomst van de Bijlmermeer." Amsterdam: SWOB 1991 (personal archive of T. van den Klinkenberg, Amsterdam).

109. Interview with Mulder.

110. See Project Office Renewal Bijlmermeer: "Vernieuwen na de ramp." Report no. 9. Amsterdam: Projectburo Vernieuwing Bijlmermeer 1992 (Archive Projectbureau Vernieuwing Bijlmermeer Amsterdam).

111. Letter from Working Group Housing and Living Environment in the local newspaper "De Nieuwe Bijlmer" (1992) (my translation). See also the letter from Hennie Bos, chair of the Independent Residents Organisation (Bos 1992).

112. Anonymous 1992.

113. Interview with Mulder. See "Vernieuwen na de ramp." Report no. 9. Amsterdam: Projectburo Vernieuwing Bijlmermeer 1992 (Archive Projectbureau Vernieuwing Bijlmermeer Amsterdam). See also Minutes Operational Staff Renewal Bijlmermeer, November 9, 1992 (Archive Projectbureau Vernieuwing Bijlmermeer Amsterdam).

114. Interview with Lambert.

115. Van Giersbergen 1997a: 33.

116. Interview with Lambert.

117. See e.g. the advisory report by the Amsterdam Council for Town Planning (ARS 1992) and the proposal to the City Council of Amsterdam, October 26, 1990. No. 891. "De toekomst van de Bijlmermeer." (personal archive of T. van den Klinkenberg, Amsterdam).

118. Westrik 1997: 54.

119. Kloos 1997: 22.

120. Interview with Pattiwael. Explaining the difficulty to change the Bijlmermeer by pointing at the existence of diverging viewpoints, is more in line with my second conceptions of obduracy that stresses the role of (opposite) dominant frames (see chapter 2).

121. Interview with Frieling.

122. Interview with Janssen.

123. Interview with Lambert.

124. "De Bijlmermeer is mijn stad. Visie voor de ruimtelijke vernieuwing van de Bijlmermeer." Rotterdam/Arnhem: Kuiper Compagnons. Bureau voor Ruimtelijke Ordening en Architectuur B.V. 1996 (obtained via Kuiper Compagnons).

125. Cited on p. 34 of van Giersbergen 1997a.

126. Interview with Bhalotra.

127. Ibid.

128. Ibid.

129. Ibid.

130. De Wagt 1996.

131. According to George Munnik, the name Bijlmer Museum was suggested by Pi de Bruijn to Paul Bos, a project manager in the G/K area, who accused these people of trying to make the Bijlmermeer a museum. Although negatively intended, the group adopted this name because it nicely captured their aim of preserving the Bijlmermeer. Between 1992 and 1994, there was a Working Group Bijlmer Museum. This working group became a foundation in order to be able to act as a legal corporation. Interview with G. Munnik, secretary of the Bijlmer Museum Foundation; Amsterdam, August 26, 1998.

132. The G/K area consists of a part of action area Ganzenhoef and a part of action area Kraaiennest.

133. Namely the following apartment buildings: Gooioord, Groeneveen, Grubbehoeve, Grunder, Kikkenstein, Kruitberg, Kleiburg, and Koningshoef.

134. Interview with Janssen.

135. See Bijlmer Museum Foundation article 2.2, April 26, 1994 (personal archive of G. Munnik, Amsterdam).

136. For the reaction of the Bijlmer Museum Foundation to Bhalotra's first sketch, see "Reactie op Bhalotra's ideeënschets Bijlmermeer." Amsterdam 1995 (personal archive of G. Munnik, Amsterdam).

137. See "Stichting Bijlmermuseum. Bewoners komen op voor de GK-buurt." Amsterdam: Stichting Bijlmermuseum 1994 (personal archive of G. Munnik, Amsterdam).

138. Interview with S. Tjoa, secretary Project Office Renewal Bijlmermeer (1994–1997), director MP-bureau (Multiculturalisation and Participation) (since 1998), Amsterdam, November 17, 1998.

139. See brochure "'Zwart-Wit' nader beschouwd. Een aanzet tot multiculturalisatie en aandeelhouderschap van de Bijlmermeer." Report no. 21. Amsterdam: Zwart Beraad en Allochtonen Breed Overleg 1997. (p. 3) (my translation) (personal archive of S. Tjoa, Amsterdam).

140. The MP office (Multiculturalisation and Participation) was established to support and monitor projects that aim at this. See brochure Black Council: "'Zwart-Wit' nader beschouwd. Een aanzet tot multiculturalisatie en aandeelhouderschap van de Bijlmermeer." Report no. 21. Amsterdam: Zwart Beraad en Allochtonen Breed Overleg 1997 (personal archive of S. Tjoa, Amsterdam).

141. See articles of the Bijlmer Museum Foundation: art. 2a+b (personal archive of G. Munnik, Amsterdam).

142. See the report of a conference on the Bijlmer Museum: "Rapportage mini-conferentie Bijlmermuseum." Amsterdam: Gemeente Amsterdam Project Management Bureau 1998 (personal archive of B. Lavell Amsterdam). Interview with van Herwaarden.

143. See a brochure issued by the Bijlmer Museum Foundation: "GK-buurt: Stedelijkheid en Natuur. Bijdrage aan brainstormsessies Projectgroepen Vernieuwing over "Bijlmermuseum, problemen oplossen binnen de oorspronkelijke ontwerpuitgangspunten." Amsterdam: Stichting Bijlmermuseum 1996, p. 2. (personal archive of G. Munnik, Amsterdam).

144. See a brochure issued by the Bijlmer Museum Foundation: "Stichting Bijlmermuseum. Bewoners komen op voor de GK-buurt." Amsterdam: Stichting Bijlmermuseum 1994 (personal archive of G. Munnik, Amsterdam).

145. See the articles of the Bijlmer Museum Foundation: art. 2.e. (personal archive of G. Munnik, Amsterdam).

146. See the articles of the Bijlmer Museum Foundation: art. 2.g. (personal archive of G. Munnik, Amsterdam).

147. Interview with Munnik.

148. Interviews with Munnik and Lavell.

149. See "Plan van aanpak herontwikkeling Groeneveen e.o.," Projectgroep Ganzenhoef, July 28, 1994, (p.3). My translation (Archive Projectbureau Vernieuwing Bijlmermeer).

150. Interview with B. Lavell, vice-project manager at Kraaiennest since 1995, Amsterdam, November 17, 1998.

151. Interview with Munnik.

152. See proposal to the district council South-East: "Voordracht inzake plan van aanpak vernieuwing K-buurt." December 17, 1996. No. 38 (personal archive of B. Lavell, Amsterdam).

153. My translation. See brochure of the Bijlmer Museum Foundation. "GK-buurt: Stedelijkheid en Natuur. Bijdrage aan brainstormsessies Projectgroepen Vernieuwing over "Bijlmermuseum, problemen oplossen binnen de oorspronkelijke ontwerpuitgangspunten." Amsterdam: Stichting Bijlmermuseum 1996. "Naschrift par. 1: Parkeren" (personal archive of G. Munnik, Amsterdam).

154. See brochure "GK-buurt: Stedelijkheid en Natuur. Bijdrage aan brainstormsessies Projectgroepen Vernieuwing over "Bijlmermuseum, problemen oplossen binnen de oor-

spronkelijke ontwerpuitgangspunten." Amsterdam: Stichting Bijlmermuseum 1996. "Naschrift par. 1: Parkeren" (personal archive of G. Munnik Amsterdam).

155. Interviews with van der Hilst and Munnik.

156. I use the term 'constructability' to denote the typical Dutch term of 'maakbaarheid'. 'Maakbaarheid' refers to the belief that social change can to a large extent be effected by government policies.

157. The term "Bijlmer believer" is commonly used in the press and in the discussions about the renewal of the Bijlmermeer to indicate those people who keep 'loving' the Bijlmermeer. Although this term has no negative meaning for me, the people who feel addressed by this term do not like it because in their view it has a pejorative connotation.

158. I will return to such strategies to preserve obduracy in chapter 5.

CHAPTER 5

1. Aibar and Bijker 1997: 12–15.

2. Klein and Kleinman 2002: 40.

3. Bos, Mik, and Versnel 1979: 13.

4. See my discussion of Harvey's work in chapter 1.

5. In June 2003, agreement was reached by the City of Maastricht, the Province of Limburg, and the national government that a tunnel will be built in Maastricht after 2007.

6. For a similar line of argumentation, linked to actor-network theory, see Williams and Edge 1996. It should be noted however that in other interpretations of the notion of embeddedness as in Harvey's work, there *is* attention for structure and power factors like the forces of capitalism.

7. Williams and Edge 1996: 890.

8. Williams 1990: 206.

9. Of course, actors may use more than one unbuilding strategy. Bakker combined the strategy of establishing historical continuity and the strategy to combine design and process interventions.

10. Ironically, Bijlmermeer residents had already shown that physical determinism does not "work": their local practices were often in radical opposition with the intentions of the designers of the Bijlmermeer.

11. Neville-Sington and Sington 1993: 37.

12. Brand 1994: 10.

13. Ibid.: 12.

14. Rossi 1982: 139.

15. Ibid.: 55.

16. Brand 1994: 11.

17. Cullingworth 1997.

18. Spaans 2000.

19. Cullingworth 1997.

20. For an analysis of cultural, historical, legal, and physical factors that shape planning systems in the UK, see Cullingworth and Nadin 2002 (1964). Cullingworth and Nadin also draw comparisons with the US and the Netherlands.

---

21. MacKenzie and Wajcman 1999: 18.

22. The idea that artifacts can be obdurate and flexible at the same time is expressed in the concept of "boundary object." See Star and Griesemer 1989.

23. For a similar argument, see Rip and Kemp 1998.

24. See e.g. Bijker 1995, 1996; Schot 1996; Rip, Misa, and Schot 1995. For a critique of these viewpoints, see Harbers 1996.

25. In this study, the InfraLab procedure is an example of such a participation model. Other recent examples are "Forum Amsterdam," "Het kan verke(e)ren" in Groningen, and "Dordt spreekt" in Dordrecht.

Sources

Interviews

Chapter 2

H. Bakker, Eindhoven, August 14, 1998

A. Bley, Utrecht, June 6, 1997

E. Bolt, Heerlen, July 9, 1997

E. Brandes, Utrecht, July 8, 1997

M. Dendermonde, Maastricht, December 10, 1997

A. Feddes, Bunnik, June 19, 1997

G. Groener, Utrecht, June 30, 1997

H. van Herwaarden, Utrecht, July 1, 1997

A. Hordijk, telephone, November 18, 1997

H. Kernkamp, Utrecht, July 8, 1997

C. Koemans, Utrecht, July 8, 1997

L. Lambo, Utrecht, April 29, 1997

A. Lambooij, Eindhoven, February 26, 1998

G. Mik, Utrecht, June 19, 1997

P. Nyst, Utrecht, July 1, 1997

J. Peters, Utrecht, December 6, 1997

D. Regenboog, Rotterdam, September 29, 1997

A. Smits, Utrecht, June 12, 1997

H. S. Yap, Den Haag, September 28, 1997

Chapter 3

A. Cremers, Beek, September 14, 1999

R. Daniëls, Maastricht, April 21, 1999

F. Hendrikx, Maastricht, February 24, 1999

J. Jamin, Maastricht, February 2, 1999

P. Jansen, Maastricht, March 4, 1999

T. Jenniskens, Maastricht, April 15, 1999

O. de Jong, Maastricht, May 7, 1998

O. Kars, Maastricht, July 15, 1999

C. Konsten, Utrecht, April 20, 1999

J. Kroon, Apeldoorn, August 18, 1999

H. Luijpers, Maastricht, April 6, 1999

A. Lutters, Maastricht, July 21, 1999

J. Magnée, Deurne, November 11, 1999

J. Nakken, Maastricht, October 28, 1999

E. Schreuders, Maastricht, April 19, 1999

J. Smeekes, Leidschendam, July 6, 1995

A. Vesseur, Maastricht, April 16, 1999

CHAPTER 4

A. Bhalotra, Rotterdam, August 18, 1998

J. Brewster, Amsterdam, July 22, 1999

P. de Bruijn, Amsterdam, August 26, 1998

D. Frieling, Delft, August 18, 1998

R. Grotendorst, Amsterdam, August 10, 1998

L. van Herwaarden, Amsterdam, July 22, 1999

S. van der Hilst, Amsterdam, July 14, 1999

R. Janssen, Amsterdam, July 31, 1998

T. van den Klinkenberg, Amsterdam, May 25, 1998

D. Lambert, Rotterdam, July 27, 1998

B. Lavell, Amsterdam, November 17, 1998

M. Mulder, Almere, August 10, 1998

G. Munnik, Amsterdam, August 26, 1998

N. Pattiwael, Amsterdam, June 4, 1998

I. Roovers, Almere, May 25, 1998

S. Tjoa, Amsterdam, November 17, 1998

ARCHIVES

CHAPTER 2

Archive of Stadsbalie, Utrecht

Archive of NS Vastgoed, Utrecht

Personal archive of L. Lambo, Utrecht

Personal archive of G. Groener, Utrecht

Personal archive of C. Koemans, Utrecht

City archive, Utrecht

CHAPTER 3

City archive, Maastricht

Archive of Stadsontwikkeling en Grondzaken, Maastricht

Archive of Rijkswaterstaat Directie Limburg, Afdeling Integraal Verkeer en Vervoer Maastricht

Archive of Bouwdienst Rijkswaterstaat, Utrecht

Archive of Sociaal Historisch Centrum, Maastricht

Personal archive of O. de Jong, Maastricht

Personal archive of J. Nakken, Utrecht

Personal archive of J. Jamin, Maastricht

CHAPTER 4

Archive of Stadsdeel Zuidoost, Amsterdam

Archive of Projectbureau Vernieuwing Bijlmermeer, Amsterdam

Personal archive of T. van den Klinkenberg, Amsterdam

Personal archive of G. Munnik, Amsterdam

Personal archive of S. Tjoa, Amsterdam

Personal archive of B. Lavell, Amsterdam

Archive of Stedelijke Woningdienst, Amsterdam

PUBLICATIONS

Aarden, M. 1997. Bouwplaats Nederland. *De Volkskrant*, March 18.

Abbate, J. 1999. Cold war and white heat: The origins and meanings of packet switching. In *The Social Shaping of Technology*, ed. D. MacKenzie and J. Wajcman, second edition. Open University Press.

Achterhuis, H. 1998. *De erfenis van de utopie*. Amsterdam: Ambo.

Aibar, E., and Bijker, W. 1997. Constructing a city: The Cerdà Plan for the extension of Barcelona. *Science, Technology, and Human Values* 22, no. 1: 3–30.

Angenot, L. 1948. Het planning-aspect. Paper presented at Autosnelwegen in en rond de stad conference, Utrecht.

Anonymous. 1973a. Hoog Catharijne (2). Uniek object. *Utrechts Nieuwsblad*, September 22.

Anonymous. 1973b. PvdA wijst procedure Hoog-Catharijne af. *De Volkskrant*, September 25.

Anonymous. 1977a. Demonstranten schrijven aan raadsleden. *De Limburger*, April 8.

Anonymous. 1977b. Moeders eisen voor kleuters veilige oversteek. *De Limburger*, April 7.

Anonymous. 1978. De Architecten: Angst voor 7 miljoen 'stedebouwkundigen' maakt tweede Hoog-Catharijne onmogelijk. *Utrechts Nieuwsblad*, September 23.

Anonymous. 1983. Omwonenden A2/E9 trekken aan de bel. Stadstraverse wordt dodenweg. *De Limburger*, February 1.

Anonymous. 1989. Argwaan bij insprekers tegenover plannenmakers van City Projekt. *Utrechts Nieuwsblad*, May 9.

Anonymous. 1991. ABP wenst geen grote doorbraak in HC. *Utrechts Nieuwsblad*, January 17.

Anonymous. 1992. Ontroering en schaamte bij Bijlmerramp. *De Nieuwe Bijlmer*, November 5.

ARS (Amsterdamse Raad voor de Stadsontwikkeling). 1992. Over de scenario's voor de Bijlmermeer. Adviesnr. 146, no. 7.

Bakker, R. 1998. Inaugural address, Technische Universiteit Eindhoven.

Battani, M., Hall, D., and Powers, R. 1997. Cultures' structures: Making meaning in the public sphere. *Theory and Society* 26: 781–812.

Berggren, H. 1956. *Maastricht in 1956*. Maastricht: Leiter-Nypels.

Bertolini, L., and Spit, T. 1998. *Cities on Rails: The Redevelopment of Railway Station Areas*. Spon.

Bijker, W. 1987. The social construction of Bakelite: Toward a theory of invention. In *The Social Construction of Technological Systems*, ed. W. Bijker et al. MIT Press.

Bijker, W. 1995a. Democratisering van de Technologische Cultuur. Inaugural address, Rijksuniversiteit Limburg, Maastricht.

Bijker, W. 1995b. *Of Bicycles, Bakelites, and Bulbs: Toward a Theory of Sociotechnical Change*. MIT Press.

Bijker, W. 1995c. Sociohistorical technology studies. In *Handbook of Science and Technology Studies*, ed. S. Jasanoff et al. Sage.

Bijker, W. 1996. Politisering van de technologische cultuur. *Kennis en methode* 20, no. 3: 294–307.

Bisscheroux, N., and Minis, S. 1997. *Architectuurgids Maastricht 1895–1995*. Maastricht: Stichting Topos, Dienst SOG Gemeente Maastricht.

Blijstra, R. 1969. *2000 jaar Utrecht: Stedebouwkundige ontwikkeling van castrum tot centrum*. Utrecht: Bruna.

Blokker, B. 1997. De kruimelstad. De openbare ruimte in de stad wordt steeds minder openbaar. *NRC Handelsblad*, June 28.

Boesenkool, J., Cornelissen, P., Huitink, H., and Stoffels, P. 1983. *Gewoon Utrecht, en hoe woont dat? 1945–1980: Van provinciestad tot betondorp*. Utrecht: Huisdrukkerij RUU de Uithof.

Bolte, W., and Meijer, J. 1981. *Van Berlage tot Bijlmer: Architektuur en stedelijke politiek*. Nijmegen: SUN.

Boomkens, R. 1998. *Een drempelwereld: Moderne ervaring en stedelijke openbaarheid*. Rotterdam: NAi.

Bos, H. 1992. Vernieuwing in nieuw licht. *De Nieuwe Bijlmer*, November 19.

Bos, K., Mik, G., and Versnel, H. 1979. Utrecht heeft weinig geleerd van vijf jaar Hoog Catharijne. *Wonen-TA/BK* 7: 12–22.

Bost, W. 1983. *Maastricht 1983*. Maastricht: Leiter-Nypels.

Brain, D. 1994. Cultural production as "society in the making": Architecture as an exemplar of the social construction of cultural artifacts. In *The Sociology of Culture*, ed. D. Crane. Blackwell.

Brakenhoff, A., Dignum, K., Wagenaar, M., and Westzaan, M. 1991. *Hoge bouw, lage status: Overheidsinvloed en bevolkingsdynamiek in de Bijlmermeer*. Amsterdam: Instituut voor Sociale Geografie.

Brand, S. 1994. *How Buildings Learn: What Happens After They're Built*. Penguin.

Bruijnzeels, K. 1960. *Maastricht in 1960*. Maastricht: Leiter-Nypels.

Brusse, P. 1997. Chaos achter de zeewering. *De Volkskrant*, January 18.

Buiter, H. 1993. *Hoog Catharijne*. Utrecht: Matrijs.

Burby, R., and Kaiser, E. 1988. How can we assess the content of urban research? *Urban Affairs Quarterly* 24, no. 1: 33–38.

Burgess, E., ed. 1925. *The Urban Community: Selected Papers from the Proceedings of the American Sociological Society*. Greenwood.

Burgess, P. 1996. Should planning history hit the road? An examination of the state of planning history in the United States. *Planning Perspectives* 11: 201–224.

Callon, M. 1986. Some elements of a sociology of translation: Domestication of the scallops and the fishermen of St. Brieuc Bay. In *Power, Action and Belief*, ed. J. Law. Routledge and Kegan Paul.

Callon, M. 1987. Society in the making: The study of technology as a tool for sociological analysis. In *The Social Construction of Technological Systems*, ed. W. Bijker et al. MIT Press.

Callon, M. 1991. Techno-economic networks and irreversibility. In *A Sociology of Monsters*, ed. J. Law. Routledge.

Callon, M. 1995. Technological conception and adoption networks: Lessons for the CTA practitioner. In *Managing Technology in Society*, ed. A. Rip et al. Pinter.

Crimson. 1997. *Re-Urb. Nieuwe plannen voor oude steden*. Rotterdam: 010.

Cullingworth, B. 1997. *Planning in the USA: Policies, Issues, and Processes*. Routledge.

Cullingworth, B., and Nadin, V. 2002 (1964). *Town and Country Planning in the UK*, thirteenth edition. Routledge.

Damoiseaux, R. 1981. *Maastricht 1981*. Maastricht: Leiter-Nypels.

Das, R., Leeflang, S., and Rothuizen, W. 1966. *Op zoek naar leefruimte. Nederland met zijn bevolkingsgroei kan best bewoonbaar blijven dankzij nieuwe technieken*. Amersfoort: Roelofs van Goor.

Daumas, M., ed. 1977. *Analyse historique de l'évolution des transports en commun dans la Région Parisienne*. Editions du CNRS.

de Boer, N., and Lambert, D. 1987. *Woonwijken: Nederlandse stedebouw 1945–1985*. Rotterdam: 010.

Deelstra, T., van Toorn, J., and Bremer, J., eds. 1972. *De straat, vorm van samenleven*. Eindhoven: Stedelijk van Abbemuseum.

de Geus, M. 1996. *Ecologische utopieën. Ecotopia's en het mileudebat*. Utrecht: Jan van Arkel.

De Grote Bosatlas. 1998. Groningen: Wolters-Noordhoff.

de Jong, R. 1972. Herovering van de straat? Sociaal-economische ontwikkeling in de negentiende en twintigste eeuw. In *De straat, vorm van samenleven*, ed. T. Deelstra et al. Eindhoven: Stedelijk van Abbemuseum.

de Rooij, A. 1994. Het Infralab-initiatief vraagt om een vernieuwde vervoersplanologie. Paper, Hoofddirectie Rijkswaterstaat, afdeling Innovatie and Synthese.

de Rouw, W. 1970. *Maastricht 1970*. Maastricht: Leiter-Nypels.

Dettingmeijer, R. 1988. Van Fockema Andreae tot renovatie van HC. In *De ideale stad*, ed. K. Jacobs and L. Smit. Centraal Museum Utrecht.

de Valk, W. 1973. De van God gegeven, in beton gegoten onwrikbaarheid van Hoog Catharijne. *De Volkskrant*, July 14.

De Volkskrant. 1997. *VOL. Het debat over de ruimtelijke inrichting van Nederland*. Zaandam: Huig.

de Wagt, W. 1996. De stedebouwer. *De Groene Amsterdammer*, July 17.

de Wit, B. 1993. Van verguisd beton naar leefbare derde-wereldstad. In *Ontwerpen voor de onmogelijke stad*, ed. R. Boomkens. Amsterdam: De Balie.

Dibbits, H. 1965. De ontwikkeling van het verkeer als factor in de ruimtelijke ontwikkeling van ons land. *Wetenschap en samenleving* 19, June/July: 77–87.

Dijkhuis, J., Ferf-van den Broeke, I., van der Maesen, L., Melger, R., and Klaren, M. 1975. *Collectieve ruimten Bijlmermeer: Analyse van een verschijnsel.* Amsterdam: Gemeentelijke Dienst Volkshuisvesting.

Dosi, G. 1982. Technological paradigms and technological trajectories: A suggested interpretation of the determinants and directions of technical change. *Research Policy* 11: 147–162.

Douma, C. 1998. *Stationsarchitectuur in Nederland 1938–1998.* Zutphen: Walburg Pers.

Ellis, C. 1996. Professional conflict over urban form: The case of urban freeways, 1930 to 1970. In *Planning the Twentieth-Century American City*, ed. M. Sies and C. Silver. Johns Hopkins University Press.

Essers, B. 1969. *Maastricht in 1969.* Maastricht: Leiter-Nypels.

Frisbie, W., and Kasarda, J. 1988. Spatial processes. In *Handbook of Sociology*, ed. N. Smelser. Sage.

Garud, R., and Karnøe, P., eds. 2001. *Path Dependence and Creation.* Erlbaum.

Ghirardo, D. 1996. *Architecture after Modernism.* Thames and Hudson.

Gicryn, T. 2000. A space for place in sociology. *Annual Review of Sociology* 26: 463–496.

Gieryn, T. 2002. What buildings do. *Theory and Society* 31: 35–74.

Gorman, M., and Carlson, W. 1990. Interpreting invention as a cognitive process: The case of Alexander Graham Bell, Thomas Edison and the telephone. *Science, Technology, and Human Values* 15, no. 2: 131–164.

Gottdiener, M., and Feagin, J. 1988. The paradigm shift in urban sociology. *Urban Affairs Quarterly* 24, no. 2: 163–187.

Graham, S., and Marvin, S. 1996. *Telecommunications and the City: Electronic Spaces, Urban Places.* Routledge.

Graham, S., and Marvin, S. 2001. *Splintering Urbanism: Networked Infrastructures, Technological Mobilities and the Urban Condition.* Routledge.

Gullberg, A., and Kaijser, A. 1998. City Building Regimes in Post-war Stockholm. TRITA-HST Working Paper 98/3, Royal Institute of Technology, Stockholm.

Guy, S., Graham, S., and Marvin, S. 1997. Splintering networks: Cities and technical networks in 1990s Britain. *Urban Studies* 34, no. 2: 191–216.

Haagsma, I. 1976. Hoog Catharijne: een grote couveuse. *NRC Handelsblad,* August 7.

Hajer, M., and Halsema, F., eds. 1997. *Land in zicht! Een cultuurpolitieke visie op de ruimtelijke inrichting.* Amsterdam: Wiardi Beckman Stichting/Bert Bakker.

Hall, P. 1988. *Cities of Tomorrow: An Intellectual History of Urban Planning and Design in the 20th Century.* Blackwell.

Hamlett, P. 2003. Technology theory and deliberative democracy. *Science, Technology, and Human Values* 28, no. 1: 112–140.

Hannerz, U. 1980. *Exploring the City: Inquiries toward an Urban Anthropology.* Columbia University Press.

Harbers, H. 1996. Politiek van de technologie. *Kennis en Methode* 20, no. 3: 308–315.

Hård, M., and Misa, T., eds. 2003. The Urban Machine: Recent Literature on European Cities in the 20th Century. www.iit.edu/~misa/toe20/urban-machine/index.html.

Hård, M., and Stippak, M. 2003. Discourses on the Modern City and Urban Technology, 1850–2000. In The Urban Machine, ed. M. Hård and T. Misa. www.iit.edu/~misa/toe20/urban-machine/index.html.

Harvey, D. 1985. *The Urbanization of Capital*. Blackwell.

Hendriks, E., and Nijland, R. 1996. De verticale vooruitgang. *De Volkskrant*, December 21.

Hommels, A. 1995. De rol van expertise bij besluitvorming over problemen rond techniekontwikkeling. Hoe worden burgers en technici "co-makers" van een weg? MA-thesis, Universiteit Maastricht.

Hommels, A. 1997. Praatgroep voor N44. Omwonenden, weggebruikers, belangengroepen en technici betrokken bij problematiek drukke weg. *De Ingenieur* 109, no. 4: 17–19.

Hommels, A. 2000. Obduracy and urban sociotechnical change: Changing Plan Hoog Catharijne. *Urban Affairs Review* 35, no. 5: 649–676.

Hofland, H. 1996. Markermeer of Markerwaard. De radicale herbouw van Nederlands is al lang in volle gang. *NRC Handelsblad*, November 23.

Hughes, T. 1983. *Networks of Power: Electrification in Western Society, 1880–1930*. Johns Hopkins University Press.

Hughes, T. 1987. The evolution of large technological systems. In *The Social Construction of Technological Systems*, ed. W. Bijker et al. MIT Press.

Hughes, T. 1988. The seamless web: Technology, science, et cetera, et cetera. In *Technology and Social Process*, ed. B. Elliott. Edinburgh University Press.

Hughes, T. 1994. Technological momentum. In *Does Technology Drive History? The Dilemma of Technological Determinism*, ed. M. Smith and L. Marx, second edition. MIT Press.

Hughes, T. 1998. *Rescuing Prometheus*. Pantheon Books.

Hughes, T., and Hughes, A. 1990. *Lewis Mumford: Public Intellectual*. Oxford University Press.

Huisman, J. 1998. De zwarte jaren zestig. *De Volkskrant*, January 3.

Huisman, J. 1999. Gescharrel in de ruimte. *Vrij Nederland*, December 11.

Huls, B., and Bomers, R. 1983. Advies inzake de Bijlmermeer. Amsterdamse Raad voor de Stedebouw.

Hulsman, B. 1999. Herzien. *NRC Handelsblad*, December 8.

Jacobs, J. 1964 (1961). *The Death and Life of Great American Cities: The Failure of Town Planning*. Penguin/Jonathan Cape.

Joerges, B. 1999. Do politics have artefacts? *Social Studies of Science* 29, no. 3: 411–431.

Johnson-McGrath, J. 1997. Who built the built environment? Artifacts, politics, and urban technology. *Technology and Culture* 38, no. 3: 690–696.

Kiers, A. 1997. Hoog Catharijne niet oogstrelend bedoeld. *De Volkskrant*, September 6.

Kitt Chappell, S. 1989. Urban ideals and the design of railroad stations. *Technology and Culture* 30: 354–375.

Klein, H., and Kleinman, D. 2002. The social construction of technology: Structural considerations. *Science, Technology, and Human Values* 27, no. 1: 28–52.

Kloos, M. 1997. A perpetual stumbling block. *Archis* 3: 22–23.

Konvitz, J. 1985. *The Urban Millennium: The City-Building Process from the Early Middle Ages to the Present*. Southern Illinois University Press.

Konvitz, J., Rose, M., and Tarr, J. A. 1990. Technology and the city. *Technology and Culture* 31, no. 2: 284–294.

Koolhaas, R. 1977. Wat betreft de Bijlmer. In *Bijlmerstrip*, ed. M. Campo et al. Technische Universiteit Delft.

Kuipers, M. 1987. *Bouwen in beton. Experimenten in de volkshuisvesting voor 1940.* Den Haag: Staatsuitgeverij.

Kune, H., and van Erkel, F. 1995. InfraLab experimenteert met interactieve planvorming. Rijkswaterstaat leert door te doen.

Lange, G. 1948. Het stedebouwkundig aspect. Paper presented at Autosnelwegen in en rond de stad, Utrecht.

La Porte, T., ed. 1991. *Social Responses to Large Technical Systems.* Kluwer.

Latour, B. 1987. *Science in Action: How to Follow Scientists and Engineers through Society.* Harvard University Press.

Latour, B. 1988. The prince for machines as well as for machinations. In *Technology and Social Process,* ed. B. Elliott. Edinburgh University Press.

Latour, B. 1996. *Aramis or the Love of Technology.* Harvard University Press.

Larson, M. 1993. *Behind the Postmodern Facade: Architectural Change in Late-Twentieth-Century America.* University of California Press.

Law, J. 1987. Technology and heterogeneous engineering: the case of portuguese expansion. In *The Social Construction of Technological Systems,* ed. W. Bijker et al. MIT Press.

Law, J. 1991. Power, discretion and strategy. In *A Sociology of Monsters,* ed. J. Law. Routledge.

Lewis, T. 1997. *Divided Highways: Building the Interstate Highways, Transforming American Life.* Viking.

Lindner, R. 1996 (1990). *The Reportage of Urban Culture: Robert Park and the Chicago School.* Cambridge University Press.

Logan, J., and Molotch, H. 1987. *Urban Fortunes: The Political Economy of Place.* University of California Press.

Luckin, B. 1991. Sites, cities, and technologies. *Journal of Urban History* 17, no. 4: 426–433.

Luijten, A. 1997. A barrel of contradictions: The dynamic history of the Bijlmermeer. *Archis* 3: 14–21.

Lynch, K. 1990 (1958). Environmental adaptability. In *City Sense and City Design*, ed. T. Banerjee and M. Southworth. MIT Press. Reprinted from *Journal of the American Institute of Planners* 24, no. 1: 16–24.

MacKenzie, D. 1995. *Knowing Machines: Essays on Technical Change*. MIT Press.

MacKenzie, D., and Wajcman, J. 1999a. Introductory essay: The social shaping of technology. In *The Social Shaping of Technology*, ed. D. MacKenzie and J. Wajcman, second edition. Open University Press.

MacKenzie, D., and Wajcman, J., eds. 1999b. *The Social Shaping of Technology*, second edition. Open University Press.

Maitland, B. 1985. *Shopping Malls: Planning and Design*. Construction Press.

Mayntz, R., and Hughes, T., eds. 1988. *The Development of Large Technical Systems*. Campus.

McKay, J. 1988. Comparative perspectives on transit in Europe and the United States, 1850–1914. In *Technology and the Rise of the Networked City in Europe and America*, ed. J. Tarr and G. Dupuy. Temple University Press.

McShane, C. 1994. *Down the Asphalt Path: The Automobile and the American City*. Columbia University Press.

Melger, R., de Haan, J., van Lammeren, L., and Teune, W. 1987. *Effectrapportage Hoogbouw Bijlmermeer*. Amsterdam: Gemeentelijke Dienst Volkshuisvesting.

Mentzel, M. 1989. *Bijlmermeer als grensverleggend ideaal. Een studie over de Amsterdamse stadsuitbreidingen*. Delftse Universitaire Pers.

Meyerson, M. 1961. Utopian traditions and the planning of cities. *Daedalus* 89, winter: 180–193.

Ministerie van Verkeer en Waterstaat. 1988. Tweede Structuurschema Verkeer en Vervoer (Deel a: beleidsvoornemen no. 20 922). Den Haag: Ministerie van Verkeer en Waterstaat.

Misa, T. 1988. How machines make history, and how historians (and others) help them to do so. *Science, Technology, and Human Values* 13, no. 3,4: 308–331.

Moehring, E. 1982. Public Works and Urban History: Recent Trends and New Directions. *Essays in Public Works History* 13: 1–60.

Moehring, E. 1990. The networked city: A Euro-American view. *Journal of Urban History* 17, no. 1: 89–97.

Monkkonen, E. 1988. *America Becomes Urban: The Development of U.S. Cities and Towns, 1780–1980.* University of California Press.

Moore, S. 2001. *Technology and Place: Sustainable Architecture and the Blueprint Farm.* University of Texas Press.

Mumford, L. 1966. *The Myth of the Machine,* volume 1: *Technics and Human Development.* Harcourt Brace Jovanovich.

Mumford, L. 1973 (1965). Utopia, the city and the machine. In *Utopias and Utopian Thought,* ed. F. Manuel, first British edition. Souvenir.

Needham, D., Kruijt, B., and Koenders, P. 1993. *Urban Land and Property Markets in the Netherlands.* UCL.

Neville-Sington, P., and Sington, D. 1993. *Paradise Dreamed: How Utopian Thinkers Have Changed the Modern World.* Bloomsbury.

N. V. Maatschappij voor Projectontwikkeling 'Empeo'. 1962. Plan Hoog Catharijne. Bijdrage tot Utrechts centrumfunktie. Utrecht: EMPEO.

N. V. Maatschappij voor Projectontwikkeling 'Empeo'. 1963. Hoog Catharijne, aanvulling. Utrecht: EMPEO.

Park, R., Burgess, E., and McKenzie, R. 1925 (1984). *The City: Suggestions for Investigation of Human Behavior in the Urban Environment.* University of Chicago Press.

Pels, D. 1997. Mixing metaphors: Politics or economics of knowledge? *Theory and Society* 26: 685–717.

Perin, C. 1977. *Everything in Its Place: Social Order and Land Use in America.* Princeton University Press.

Peters, P. 1998. De smalle marges van de politiek. In *Cultuur en Mobiliteit*, ed. H. Achterhuis and B. Elzen. Den Haag: Rathenau Instituut.

Pinch, T. 2001. Why do you go to a piano store to buy a synthesizer? Path dependence and the social construction of technology. In *Path Dependence and Creation*, ed. R. Garud and P. Karnøe. Erlbaum.

Pinch, T., and Bijker, W. 1984. The social construction of facts and artifacts: or how the sociology of science and the sociology of technology might benefit each other. *Social Studies of Science* 14, no. 3: 399–441.

Pinch, T., and Bijker, W. 1987. The social construction of facts and artifacts: or how the sociology of science and the sociology of technology might benefit each other. In *The Social Construction of Technological Systems*, ed. W. Bijker et al. MIT Press.

Pol, M., and Nägele, R. 1996. Verslag Open Planproces Rijksweg 2 Passage Maastricht (no. TT96-31): Traffic Test bv. Instituut voor onderzoek en beleidsadvisering op gebied van verkeer en vervoer.

Priemus, H. 1986. *"The Spirit of St. Louis": De neergang van Pruitt-Igoe.* Delftse Universitaire Pers.

Provoost, M. 1996. *Asfalt. Automobiliteit in de Rotterdamse stedebouw.* Rotterdam: 010.

Rip, A., and Kemp, R. 1998. Technological change. In *Human Choice and Climate Change*, ed. S. Rayner and E. Malone. Batelle.

Rip, A., Misa, T., and Schot, J., eds. 1995. *Managing Technology in Society: The Approach of Constructive Technology Assessment*. Pinter.

Rose, M., and Clark, J. 1979. Light, heat, and power: Energy choices in Kansas City, Wichita, and Denver, 1900–1935. *Journal of Urban History* 5, no. 3: 340–364.

Rosen, C. 1986. Infrastructural improvement in nineteenth-century cities: A conceptual framework and cases. *Journal of Urban History* 12, no. 3: 211–256.

Rosen, C. 1989. Review of *The City and Technology*. *Technology and Culture* 30, no. 4: 1070–1072.

Rossi, A. 1982. *The Architecture of the City*, sixth edition. MIT Press.

Sansone, L. 1992. *Schitteren in de schaduw. Overlevingsstrategieën, subcultuur en etniciteit van Creoolse jongeren uit de lagere klasse in Amsterdam 1981–1990*. Amsterdam: Het Spinhuis.

Schmucki, B. 2003. The city and urban transport: A bibliographic overview. In The Urban Machine, ed. M. Hård and T. Misa. www.iit.edu/~misa/toe20/urban-machine/index.html.

Schot, J. 1996. De inzet van constructief technology assessment. *Kennis en methode* 20, no. 3: 265–293.

Schuman, T., and Sclar, E. 1996. The impact of ideology on American town planning: From the Garden City to Battery Park City. In *Planning the Twentieth-Century American City*, ed. M. Sies and C. Silver. Johns Hopkins University Press.

Schuyt, K., and Taverne, E. 2000. *Nederlandse cultuur in Europese context. 1950. Welvaart in zwart-wit*. Den Haag: Sdu.

Schwab, W. 1992. *The Sociology of Cities*. Prentice-Hall.

Scott, J. 1998. *Seeing Like a State: How Certain Schemes to Improve the Human Condition Have Failed.* Yale University Press.

Sies, M., and Silver, C. 1996a. Introduction: The history of planning history. In *Planning the Twentieth-Century American City*, ed. M. Sies and C. Silver. Johns Hopkins University Press.

Sies, M., and Silver, C., eds. 1996b. *Planning the Twentieth-Century American City.* Johns Hopkins University Press.

Spaans, M. 2000. Realisatie van stedelijke revitaliseringsprojecten. Een internationale vergelijking. Ph.D. thesis, Technische Universiteit Delft.

Star, S., and Griesemer, J. 1989. Institutional ecology, "translations," and boundary objects: Amateurs and professionals in Berkeley's Museum of Vertebrate Zoology, 1907–39. *Social Studies of Science* 19: 387–420.

Staudenmaier, J. 1985. *Technology's Storytellers: Reweaving the Human Fabric.* MIT Press.

Stichting Wijkopbouworgaan Bijlmermeer. 1980. Van de Bijlmer méér maken. Een deltaplan voor de Bijlmermeer. Amsterdam: SWOB.

Stoppelenburg, P. 1972. Nieuwe stedebouw als maatschappelijk verschijnsel. In *De straat, vorm van samenleven*, ed. T. Deelstra et al. Eindhoven: Stedelijk van Abbemuseum.

Summerton, J. 1992. District Heating Comes to Town. The Social Shaping of an Energy System. Ph.D. thesis, Linköping University.

Summerton, J., ed. 1994. *Changing Large Technical Systems.* Westview.

Sutcliffe, A. 1988. Street transport in the second half of the nineteenth century: Mechanization delayed? In *Technology and the Rise of the Networked City in Europe and America*, ed. J. Tarr and G. Dupuy. Temple University Press.

Tarr, J. 1984. The evolution of the urban infrastructure in the nineteenth and twentieth centuries. In *Perspectives on Urban Infrastructure*, ed. R. Hanson. National Academy Press.

Tarr, J., and Dupuy, G. 1988a. Preface. In *Technology and the Rise of the Networked City in Europe and America*, ed. J. Tarr and G. Dupuy. Temple University Press.

Tarr, J., and Dupuy, G., eds. 1988b. *Technology and the Rise of the Networked City in Europe and America*. Temple University Press.

Tarr, J., and Konvitz, J. 1987. Patterns in the development of the urban infrastructure. *American Urbanism* 125: 195–226.

Taverne, E. 1983. Architects without architecture. Architectuurdiscussie in Nederland 1940–1980. In *Architectuur en Planning: Nederland 1940–1980*, ed. U. Barbieri. Rotterdam: 010.

ter Horst, J., Meyer, H., and de Vries, A. 1991. *Sleutelen aan de Bijlmer. Interpretaties*. Delft: Faculteit der Bouwkunde.

Thewissen, M. 1958. *Maastricht in 1958*. Maastricht: Leiter-Nypels.

Tod, I., and Wheeler, M. 1979. *Utopia. Wereldhervormers tussen werkelijkheid en fantasie*. Haarlem: De Haan.

Trefil, J. 1994. *A Scientist in the City*. Doubleday.

Truffer, B., and Dürrenberger, G. 1997. Outsider initiatives in the reconstruction of the car: The case of lightweight vehicle milieus in Switzerland. *Science, Technology, and Human Values* 22, no. 2: 207–234.

Vance, J. 1977. *This Scene of Man: The Role and Structure of the City in the Geography of Western Civilization*. Harper's College Press.

van der Cammen, H., and de Klerk, L. 1993. *Ruimtelijke ordening: De ontwikkelingsgang van de ruimtelijke ordening in Nederland*. Utrecht: Uitgeverij Het Spectrum.

van der Woud, A. 1983. *Het Nieuwe Bouwen*. Delft University Press.

van de Venne, J. 1958. Een verkeersgeleidingsplan voor de Maastrichtse city. Voordracht door de directeur van openbare werken gehouden voor de gemeenteraad op 24 november 1958 (lecture).

van de Venne, J. 1959. Verkeersplannen in Maastricht. *Publieke Werken* 27, no. 3: 4–10.

van de Venne, J. 1962. Maastricht-stad in een keurslijf. *Publieke Werken* 30, April: 46–56.

van de Venne, J. 1964a. *Maastricht, een visie op de toekomst*. Gemeente Maastricht.

van de Venne, J. 1964b. Modern Town Planning in Maastricht: A History of and Introduction to the Practice of Town Planning in Maastricht. Lecture at University of Newcastle-upon-Tyne, April 10, 1964.

van Esschoten, P., and Kragten, R. 1987a. Openbare weg lijkt hier op de binnenkant van een warenhuis. *Utrechts Nieuwsblad*, January 17.

van Esschoten, P., and Kragten, R. 1987b. Utrecht wil aansluiten bij plannen NS en Jaarbeurs. *Utrechts Nieuwsblad*, January 8.

van Giersbergen, M. 1997a. Green strips, white strips. *Archis* 3: 32–35.

van Giersbergen, M. 1997b. Working on the infrastructure. *Archis* 3: 26–28.

van Zundert, J. 1996. *Het bestemmingsplan. Een juridisch-bestuurlijke inleiding in de ruimtelijke ordening*, eighth edition. Alphen aan den Rijn: Samson H. D. Tjeenk Willink.

Verhagen, E. 1987. *Van Bijlmermeerpolder tot Amsterdam Zuidoost*. Den Haag: Sdu.

Wachs, M., and Crawford, M., eds. 1991. *The Car and the City: The Automobile, the Built Environment, and Daily Urban Life*. University of Michigan Press.

Werkgroep Wonen en Woonomgeving Bijlmermeer. 1992. Maakt de ramp verschil? *De Nieuwe Bijlmer*, November 5.

Westrik, J. 1997. Enthusiasms. New urban plans for the Bijlmermeer. *Archis* 3: 49–55.

Williams, R. 1990. *Notes on the Underground: An Essay on Technology, Society, and the Imagination.* MIT Press.

Williams, R., and Edge, D. 1996. The social shaping of technology. *Research Policy* 25: 865–899.

Winner, L. 1999 (1980). Do artifacts have politics? In *The Social Shaping of Technology*, ed. D. MacKenzie and J. Wajcman, second edition. Open University Press.

Wyatt, S. 1998. Technology's Arrow. Developing Information Networks for Public Administration in Britain and the United States. Ph.D. thesis, University of Maastricht.

Yap, H. 2000. *De stad als uitdaging. Politiek, planning en praktijk van de stedenbouw.* Rotterdam: NAi.

Kathryn Henderson, *On Line and On Paper: Visual Representations, Visual Culture, and Computer Graphics in Design Engineering*

Anique Hommels, *Unbuilding Cities: Obduracy in Urban Sociotechnical Change*

David Kaiser, editor, *Pedagogy and the Practice of Science: Historical and Contemporary Perspectives*

Peter Keating and Alberto Cambrosio, *Biomedical Platforms: Reproducing the Normal and the Pathological in Late-Twentieth-Century Medicine*

Eda Kranakis, *Constructing a Bridge: An Exploration of Engineering Culture, Design, and Research in Nineteenth-Century France and America*

Pamela E. Mack, *Viewing the Earth: The Social Construction of the Landsat Satellite System*

Donald MacKenzie, *Inventing Accuracy: A Historical Sociology of Nuclear Missile Guidance*

Donald MacKenzie, *Knowing Machines: Essays on Technical Change*

Donald MacKenzie, *Mechanizing Proof: Computing, Risk, and Trust*

Maggie Mort, *Building the Trident Network: A Study of the Enrolment of People, Knowledge, and Machines*

Nelly Oudshoorn and Trevor Pinch, editors, *How Users Matter: The Co-Construction of Users and Technologies*

Paul Rosen, *Framing Production: Technology, Culture, and Change in the British Bicycle Industry*

Susanne K. Schmidt and Raymund Werle, *Coordinating Technology: Studies in the International Standardization of Telecommunications*

Charis Thompson, *Making Parents: The Ontological Choreography of Reproductive Technology*

Dominque Vinck, editor, *Everyday Engineering: An Ethnography of Design and Innovation*

Index